David Thomas Smith

The Philosophy of Memory and Other Essays

Consisting of Articles on the Philosophy of Emphasis, the Functions of the Fluid

Wedge, the Birth of a Planet, the Laws of Riverflow

David Thomas Smith

The Philosophy of Memory and Other Essays
Consisting of Articles on the Philosophy of Emphasis, the Functions of the Fluid Wedge, the Birth of a Planet, the Laws of Riverflow

ISBN/EAN: 9783337237011

Printed in Europe, USA, Canada, Australia, Japan

Cover: Foto ©Thomas Meinert / pixelio.de

More available books at **www.hansebooks.com**

THE

PHILOSOPHY OF MEMORY

And Other Essays

Consisting of Articles on

THE PHILOSOPHY OF EMPHASIS
THE FUNCTIONS OF THE FLUID WEDGE
THE BIRTH OF A PLANET
THE LAWS OF RIVERFLOW

BY

D. T. SMITH, M.D.

Lecturer on Medical Jurisprudence in the University of Louisville

Nil tam difficile est, quin quærendo investigari possit

LOUISVILLE, KY.

Press of John P. Morton & Company

1899

To

Colonel Reuben T. Durrett, A.M., LL.D.

WHOSE GENEROUS CARE IN FOSTERING THE FILSON CLUB, ENCOURAGING
LITERATURE IN KENTUCKY, AND PROMOTING RESEARCH INTO THE
HISTORY OF THE STATE WILL EVER ENTITLE HIM TO
AN UNFADING PAGE IN HER LEGENDS, THIS
LITTLE VOLUME IS AFFECTION-
ATELY INSCRIBED BY
HIS FRIEND

The Author.

PREFACE.

In this collection of essays is embraced the result of many years of thought and observation on the part of the author, and it has seemed to him that they might in some degree prove worthy of preservation.

If any of them shall prove true, the importance of the matters to which they relate entitles them to endure.

If it shall develop that none of them is true, they may still serve to point out the direction of unsuccessful endeavor, and thus save some future investigator from a like vain waste of time and effort.

If perchance a perusal of these reveries shall afford those who take interest in such studies a pleasure comparable to that the author has experienced in their indulgence, he will not altogether have failed of reward.

THE PHILOSOPHY OF MEMORY

— OR —

THE RELATION OF MIND TO THE COMMON FORCE

INTRODUCTORY.

The theme of this essay has been a subject of inquiry on the part of the writer for a period of more than thirty years.

It first took written form about the year 1874, in a communication to the New York Sun, in an effort to explain the remarkable phenomena manifested in connection with Miss Mollie Fancher, of Brooklyn, N. Y. Later, in 1878, the subject was elaborated at greater length, published in the American Practitioner and News, and extensively circulated in the form of reprints.

In that article the position was distinctly taken that there probably can be no motion which, on final analysis, would not be found to be vibratory. It was two years later that Hertz demonstrated the vibratory nature of electricity. An effort is here made to prove the vibratory character of mass. motion.

As regards the great vital principles involved in the discussion, the reader must be left to judge whether or not progress has been made in their elucidation.

That all difficulties have been removed, or even all obstacles surmounted, the writer would be only too glad to feel assured.

But of this he does feel convinced, that whether or not the solution of the interesting and weighty problems has been reached, and in fullness it certainly has not, so many telling analogies have been found to apply in the investigation that it can not well be otherwise than that the truth has been not remotely paralleled.

THE PHILOSOPHY OF MEMORY

Or the Relation of Mind to the
Common Force.

UNTIL a very recent period investigators sought
to gain an understanding of the nature and
operations of the mind through the interrogation of
consciousness.

To this task many of the most astute intellects of
all times have addressed themselves, and there is
every reason to believe that practically every thing
is known that is ascertainable by that method. This
is quite apparent from the barrenness of nearly all
recent efforts in that direction ; for, undertake it
who might, nothing of value for more than a generation has in that way been added to the body of
mental science.

Without ignoring the valuable assistance supplied
by this method and the treasures it had yielded,
but recognizing that the mine was practically exhausted to it, the world of science welcomed the
doctrine of evolution as offering a far more prom-
ising outcome, and as affording the only light that
had been thrown into many obscure recesses hitherto
almost wholly unexplored.

More recently still a new school of psychologists
has arisen, or rather the followers of both the other
methods have resorted to the expedient of seeking

by direct experiment to reach the desired goal. But these, too, are clearly destined to fall short of the solution of the difficult problem. There remains, then, recourse to the employment of analogy, which, though making no promise of certainty, gives earnest of great helpfulness.

But deeply interesting as are all questions relating to the phenomena of mind, there are others still more interesting arising out of the inquiry into the nature and source of these manifestations. These relate to the principle of life, the tree of which mental manifestations are but the fruit, or rather but the foliage.

What is life? What is the soul? These are questions that every thinking person asks, and which as yet no one has ever been able to answer with any sort of satisfaction either to himself or the discriminating inquirer.

The essence or the substance that lies at the foundation of the phenomena of life and mind, it is not a venture to say, will never be fully grasped by human understanding. What light is shed upon it must be gathered from deduction and analogy chiefly, and that often of the most remote and uncertain character. Even after light has been gathered from every possible source, we may still be able to discern only the first glimmerings of the dawn.

As preliminary to any intelligent discussion, let us measurably determine what we are to understand by the terms to be employed. We must first, then, as otherwise all our efforts would be futile, briefly consider wherein lies the difference between the class of beings possessed of intelligence and that without, between the living and the non-living, the organic and the inorganic.

Broadly speaking, it is impossible to lay down any consistent rules, to draw any hard and fast line, between man and lower living forms, and to say that the one is distinctly marked thereby from the other. Passing down the scale, we find men and lower animal forms merging the one into the other almost insensibly ; and then animals into plants, until we reach the very lowest order of life. Here for the first time, between the organic and inorganic, do we find a distinct and unmistakable line of demarkation ; and that line is constituted by the limit of the capacity for reproduction or multiplication.

On one side of the line all are favored with the power of reproducing their kind ; on the other, none is so favored. In order to realize the completeness of the distinction marked by this line, it will be necessary briefly to consider the nature of the common forms of force, and then compare some of the principles which govern them with some of the laws which seem to govern the supposed vital force.

CHARACTERISTICS OF FORCE.

The common forms of force, as manifested in light and heat, chemical affinity, electricity, magnetism, the Roentgen rays, and motion in mass, are all known to be convertible without loss or gain, the one into the other, and no addition is ever made by such conversion to the total of energy at any time in existence, and no energy is ever either created or destroyed.

Thus a given quantity of coal when consumed will produce, by the action of the chemical affinity of its elements, a definite measure of heat or light. This heat may be transformed into electricity or magnetism, and each in turn be employed in imparting motion to a mass or the performance of various kinds of work.

But the heat, light, electricity, magnetism, and the mass motion have not in any true sense been produced by the chemical affinity. Chemical affinity has simply been changed or transformed in turn into each of these manifestations of energy. And after carrying the energy through all possible forms and then back to the one begun with, if we gather up the waste we shall have exactly the quantity with which we began.

Let a cannon ball be rolled at the rate of thirty miles an hour against another of equal weight that is still, and the still one may move away at the rate of thirty miles an hour ; but the first one will have become still. If two such balls be struck by the one rolling, they will move only fifteen miles an hour, and so on for similar cases. Only so much force can be exerted as constituted the motion of the original ball.

VITAL FORCE.

But the case is quite different with the vital energy or the principle known as life. An acorn planted in the earth grows into a giant oak, and every year the oak bears hundreds of thousands of such acorns as that from which it was itself produced. Each of these in turn may produce another oak like the parent, until in this way millions and millions of trees will have been produced, every one identical in character with the original, or parent oak.

Countless millions of tons of matter have been arranged into the form of trees, apparently in opposition to the laws of gravity, and certainly in opposition to the ordinary laws of decay. The same is true for any form of organic life whatever that may be chosen for experiment or observation.

This power of multiplication is then the one thing in which the organic demonstrably differs from the inorganic, the living from the dead. In this, truly, no man "can tell the difference between the man that walketh upright and the beast that goeth downward toward the earth;" nor, for that matter, between man and the plants of the field.

SOUL THE PRINCIPLE OF REPRODUCTION.

This principle, then, which determines multiplication, which presides over reproduction, we may denominate the soul or life. And it is not believed that there is any process of reasoning that will give man a soul that will not accord a soul to every living thing, whether animal or plant.

Mind, in so far as it can be surveyed, is merely an accident of the soul, merely one of its fruits, as the body is one, and no more identical with it than the pulp of the peach is with the mysterious essence that determines the production of the tree after its kind.

Now if there is a determining principle of life in
every plant and in every animal, that principle must
exist in unlimited extent in nature, or else it must
be capable of infinite division without impairment.
If the force employed in vital manifestations is a
modification of the common forms of natural force,
it is a modification like no other that we know. If,
on the other hand, it is of a nature different from the
common forms of force, if it is a separate and peculiar force, then we must suppose that there exists
about us in nature an unlimited store of this life
force or energy out of which souls may be formed.

Say, for example, we take the case of man, in
deference to those who profess to believe that only
man has a soul : we find ourselves soon at a loss to
know how the descendants of the first man, even in the
next generation, could have been possessed of a soul.

Let us assume that the man of each generation
shall have ten children out of possible hundreds.
The first of a line would, after no great lapse of
time, have millions of descendants, every one as
certainly possessed of a soul as himself. Likewise,
if we take the woman instead of the man. That is
to say, the original soul must have been divided and
subdivided, again and again, into millions and bill-
ions of souls, while still leaving the ancestral soul
untouched in its integrity. For, after the last of the
children in each family is born, the parents still to
all seeming have the same souls with which they
were born, wholly unimpaired.

That is, the parents live on long after the off-
spring has been produced, with no sign or symptom

to show that the parental soul has been in the least diminished. Whoever can believe this to be the case; whoever can believe that any thing conceivable can give off of its potency unendingly without diminution, need not trouble himself with science or logic; he can find mental repose in believing whatsoever most pleases, and save himself all the labor of investigation.

We have, then, three alternatives of the origin of souls or the vital individualized energy :

First : A soul is specially created for each living being, animal or vegetable, or at least for each human being, if any will have it so.

Second : The vital or soul energy or essence is but a transformation of the forms of energy common in the natural world.

Third : There is in nature a store of the particular energy out of which the soul is formed.

The notion of the special creation of a soul for each individual that comes into existence I must leave for those who think it worthy of entertainment. The fact that the life energy, with its apparent susceptibility of infinite multiplication, has the power of working contrary to and in opposition to the modes of the known force-forms in nature, makes it inconceivable that it can be a form of the common energy.

It would follow, then, that there exists in nature, somehow bound up or mingled with its coarser elements, a soul-material or essence of life that is drawn upon for the life principle of each new being that comes into existence.

IS ALL LIFE ONE?

Has this principle a separate form for every different grade of life, or does there exist only one principle which is modified for each kind of living organization, or even for each individual?

A comparison of the various forms in which life presents itself leads almost unavoidably to the conclusion that there is a close relationship, if not substantial identity, between the life principle in man and that of all lower forms of living beings.

The cases in which they respond in a similar manner to modifying conditions are very numerous.

In the process of growth all organic forms appropriate and employ practically the same chemical elements, and especially is this true for such parts as may be devoted to nutrition, either of their own tissues or those of others. Life, in both plants and animals, affords protection against decay. Thus the tree may grow for a thousand years, resisting decay, even at the heart, where apparently no cells capable of destroying microbes may have existed for many centuries; and yet as soon as it has been deprived of life, it at once, under a continuance of the same conditions, begins to decay.

In any forest during the growing season may be seen a variety of reactions on the part of the trees against injuries by different insects. Thus one kind of insect may sting a leaf or bud on certain varieties of trees, at the same time depositing an egg, and the wounded part will at once react by producing a "vinegar" ball or excrescence filled with acetic acid, and in which the egg is inclosed.

Around the egg of another insect will be developed a nutgall in which the young will be hatched. The egg and sting of still another insect will cause the twig or leaf of another tree to react with a puff ball, and so on, until a great variety of such excrescences are produced ; but a definite kind of growth will take place in the reaction against each particular character of injury.

And it is not improbable that along with these excrescences there is developed in each case a distinct kind of antitoxin to arrest the spread of the poison the sting of the insect has conveyed.

It is known that the plant, like the animal, reacts against injuries, and that a greater vigor of growth will be shown in their neighborhood. Even the repair material usually employed, like that in animals, is of a substance of an exceptionally low order of vitality, namely, scar tissues, such as nature uses for mending purposes in both classes.

Wonderful to relate, even a tree or plant and its fruit as well will, when injured, suffer a real inflammation with rise of temperature, easily shown by the thermometer ! They have a true fever.

There are species of plants that entrap insects and use them for food, secreting a kind of gastric juice and digesting them in the truest sense. Indeed, in man, as in all other animals, the final work of digestion is done by the leukocyte, which is closely related to the protoplasm in the cell structures of trees.

Reproduction in both is carried on in a way that is closely analogous, the male and female elements of flowers often making apparently consciously intelligent efforts to meet each other. A vine whose ancestors for untold ages have been climbing trees by means of tendrils will, if placed by a stone wall which the tendrils can not grasp, alter these into suckers and by their use cling on to the smooth stones for support. The same albuminoid protoplasm is the prime factor in the growth of both plants and animals.

Sensibility is a characteristic of all animals, but to a certain extent is possessed by plants also. The power of changing inorganic into organic material is a vegetable function in the main, but one that is possessed in some degree also by certain lower animals. In short, animal and vegetable life are so linked together by intermediate transitional forms that no man can positively tell where one ends and the other begins.

If, then, there is a vital energy, a peculiar soul substance, it is in all likelihood closely related in its nature in all things that multiply, and is modified in some unknown way for each species of organism.

THE NATURE OF MIND.

Having endeavored to set forth in an intelligible way the nature of the vital force or the soul element, we may now pass to a consideration of the nature of mind.

By many this has been regarded as a separate entity, the immortal part of man, an idea we have preferred to attempt to convey by the term soul. Mind is defined as that which perceives, thinks, feels, judges, desires, and wills. If mind, then, is to be defined as such only when in action, what is to be said of it in the yet unborn, the comatose, the profoundly sleeping? If, however, we consider mind as nothing more nor less than a function, no other than a recognition by consciousness of the play of certain physical forces on the soul and the soul's reaction in response to them, we give it a character that does not necessitate the existence of two separate, intangible entities in the same individual.

We may then consider the vital force that builds up the new being after certain definite and logical patterns as the entity which, when rightly played upon by certain forms of the common force of nature, or it may be when rightly ministered to by them, gives out the phenomena we call mental in the same orderly way that it builds up the bodily structure of the individual. It has been poetically said that creatures are the thoughts of God. But under this postulate it may be literally said that all organic forms, as well as ideas or mental images, are thought forms.

The highest and most perfect examples of soul work are done without consciousness, done before consciousness passes on them. In an intellectual way, to him who is known as the genius comes the richest product of intellectual activity, ready formed, and often just in all its logical relations. And if we but closely observe we will find that the great mass of our thoughts, well up into consciousness already formed, spring up out of the brain's secret chambers as plants out of the ground.

Who can trace the wonderful steps in the development of the embryo and doubt that an indwelling intelligence unknown to consciousness directs the consecutive changes as evidently as conscious or unconscious thoughts do the literary gem, whether the finished poem or the symmetrical and harmonious oration.

THOUGHTS AND THINGS.

We may perceive something like the same logic of movement, the same sequence of development, and their resultant groupings in inorganic elements ; and what is more, they present themselves almost invariably in forms agreeable to our minds. The well-known example of the formation of definite figures by placing dust particles on a tense membrane and setting it in vibration by means of musical notes is a case in point ; likewise the arrangement of particles of iron in the magnetic field. Every one has noticed how frost flowers are formed on window panes when the weather is cold. Their harmony seems as complete and their proportion as apt as if they were intelligently designed.

The lesson to be drawn from these facts is that there runs through nature a tendency to analogous groupings of the molecules of matter, indicating the tendencies of force, and that these groupings find in our thoughts a response and an approval which goes to show that the laws of thought and those of the tendencies of vital or soul force, as well as the natural forces, are similar ; that the flower appears beautiful to us because it is constructed on a plan that finds a response in the constitution of our minds; and that it is such as we would have created it had we creative powers. It appears beautiful because the thought is akin to the thing.

Furthermore, it is highly probable that there are certain susceptibilities and tendencies in particular kinds of atoms or molecules of matter that make them responsive to the affinities of vital forces, otherwise they could not be built up into organic forms. We find certain kinds of material atoms, for instance, responding to and attracted by the magnet, yet we know that the atoms thus affected must have just as much affinity for the magnet as the magnet has for them.

Only a small part of the many kinds of chemical elements is capable of being built up into animal and vegetable forms, and we may rationally conclude that only these elements or the forces that cluster around them have such affinity for the vital force as will allow it to use them.

Oxygen, hydrogen, carbon, nitrogen, sulphur, iron, and phosphorus mostly constitute the bioplasm of animals and plants alike.

But really the conditions do not seem fully satisfied except by the supposition that there exists in nature a peculiar class of vital atoms — atoms around which gathers the vital force, as the common force or energy gathers about the coarser atoms of ordinary matter, and, furthermore, that these are endowed with separate sexual tendencies.

CHARACTERISTICS OF FORCE.

Force is defined as the measure of the tendency of energy to transform itself. It might be defined more simply as that which produces motion or pressure, but motion itself is force. All forms of force have now come to be regarded as interchangeable into each other, being simply different modes of motion.

It is probably not too much to claim, also, that all force or all manifestations of energy can in the final analysis be rightly regarded as resolvable into undulations. This seems to be implied in the assumption that action and reaction are equal. In most of the manifestations of energy this fact has been demonstrated.

In others, though assumed, it is not only not demonstrated but difficult to comprehend. It is difficult to realize vibrations in stored electricity, in chemical affinity, in gravity, or in the motion of a mass, as of a cannon ball or a planet. But when these forces are changed into other forms they become vibratory, or rather manifest vibration ; as when, for instance, a cannon ball strikes the steel armor of a ship and its mass motion is changed into heat, light, and electricity.

When the powder in the cannon was exploded the chemical affinity of its elements became heat, and expanded the gas thereby produced, which set the ball in motion. The vibrations of heat became the motion of the ball, and were seemingly lost, or disappeared. But when the ball struck the ship's armor some of its molecules were driven too close together and they sprung apart, and then back and forth, thus reviving the vibrations the ball started with. Again, there were other parts of the ball at the sides that were stopped before reaching the plates, and the atoms of these parts were strained in the tendency to keep moving on, and then, being pulled back by cohesion, were set to vibrating.

Thus when we come to fully understand, probably all motion and all force can be resolved into undulations. But can we conceive that the motion of the ball in mass is vibratory; can we translate it into vibrations?

Take a dozen billiard balls and let them swing as pendulums in a line. Now let them at the point of highest speed strike all at once, and with a force equal to their momentum, a bar which, after traversing one foot, shall in turn strike all at once a dozen other balls of equal weight hanging in line.

Now the first dozen will become still, and the second dozen will be set to vibrating just as the first were vibrating before. But the bar, while moving, was seemingly not vibrating.

So the powder, when exploded in the cannon, began changing its chemical affinity into the form of vibrations known as heat. These vibrations began to be communicated to the atoms of the ball, and a majority of them in the only direction the ball could move, that is, toward the mouth of the cannon. Now so many of these atoms entered at the same time and in the same direction into the maximum positive elongation, that is, the swing or the excursion away from the powder, that the ball was carried away in mass. That is to say, the vibration of a majority of the atoms in one direction became motion of the mass in that direction. By this mass movement the vibrations were stopped when only half made, that is, when only half a period had been completed. The other half will be completed when the ball strikes the armor of a ship or any thing else and has its motion arrested ; or never, if its motion is never arrested.

VIBRATIONS HIDDEN IN MASS MOTION.

In the mass motion of the ball is therefore hidden away one half of every heat and every light vibration that will be developed when it is arrested. Even the earth is carried through space, its motion being one half of all the heat and light vibrations that would appear if it should strike the sun or the largest fixed star. So then, probably, there is no movement whatever that is not resolvable into vibrations and that would not be so resolved if the capacity of our minds were equal to a sufficiently refined analysis. It might be added that if all the atomic vibrations due to the explosion of the powder in the cannon were to take place in the one direction, at the same time, and could be taken at the same phase of their period, the ball would move away with nearly the speed of light.

The chief aim in this investigation of the nature of force has been to lead up to a just understanding of the extent and character of undulatory or vibratory motion, for upon this is to be based the attempted solution herein offered of some of the most interesting of the problems of mind.

What needs now to be proved is that it is characteristic of all forms of undulatory or vibratory motion that with time and distance the waves or vibrations increase in amplitude and decrease in intensity; and that this is probably the rule with the vibrations that result in the phenomena of mind.

This is easy to observe in the case of waves on bodies of water. Just behind the wheels of a steamer the waves are high and rapidly follow each other. As they go farther away they become longer and slower, and before disappearing they become widely separated, gentle swells. This has been found true also with earthquake shocks. Near the site of the disturbance the vibrations will be found rapid and violent. Farther away fewer will pass a particular station in a given length of time, and these will be found much gentler in character.

We have seen that in water waves, as in earth tremors, increase of amplitude keeps pace with decrease of intensity as the wave progresses. With sound waves, however, this is not by any means so obvious, for, as in the case of light, the wave length is said to be not increased with the progress of sound, and not to grow slower. The pitch or vibration frequency remains the same, whatever the distance the sound is heard. The distance over which the sound will pass in a given time, say one second, is constant for the same medium at the same temperature.

LIGHT VIBRATIONS CONSTANT.

We shall find in the case of light that all vibrations are believed to travel at the same rate, but with sound the long vibrations travel faster than the shorter ones, and a loud sound that consists of long vibrations will travel faster than a low one. So much is this the case that in the polar regions the report of a gun may first be heard by a distant observer and afterward the command of the officer to fire.

The length of a sound wave will also be diminished while passing from a rarer medium into a denser one. But in all this we have found nothing by which we may indicate the distance the sound has come. What is it, then, in the nature of sound that enables us to distinguish whether it comes from a point nearby or one farther away? As long as it moves through the same medium we have learned that there is no change in the speed of sound, which is the product of the wave length by the vibration frequency, and none in the pitch, which directly relates to the wave length ; and yet a distinct characteristic is imparted to sound by distance.

There remains, then, so far as we now can ascertain, only an undefined quality of sound by which we can distinguish whether it comes from a near point or one farther away. Quality is the name given to the impression produced on the ear by sound when it is complex ; and it may be that such complexity takes on a different character at a distant point from the source of the sound and at a near point ; but this remains to be demonstrated.

Still for all this, every one knows that as a rule there is a peculiar quality attached to sound that enables him to form some notion, usually a fairly correct one, as to the distance it has come. Furthermore, that quality can be voluntarily imitated, so that the sound of a voice may be made to seem to come from a distance, even when it is near.

One who has listened to the hunter's horn, or the baying of the hounds in the chase, knows that independently of their moving, independently of any lengthening or shortening of vibrations depending upon their coming nearer or going farther away, he can form a fair judgment as to whether the chase approaches or recedes. In the battle, the share that each rifle contributes to the rattling noise is instantaneous, the length of sound wave does not depend upon the movement of the gun farther or nearer while the sound is in the making, and yet the general, from his distant view point, can tell from the character of the sound whether victory or defeat is the promise ; can tell who advances and who retreats. Can it be that the slight peculiarity we call quality or complexity is the only thing upon which our judgment is formed, the only mark upon which we can base a distinction ?

The waves of light in some respects resemble water waves, and in others the waves of sound. The transversal waves of light resemble the circular waves of water, since in both the wave consists of a portion raised above the normal level and called the crest, and of a portion depressed below it, called the trough. But instead of the waves of light gain-

ing in amplitude and losing in intensity as do those
of water, the teaching is that they remain the same
in their progress through all distance.

The speed *in vacuo* of all light waves is believed
to be the same, no matter what the wave length
may be. If this were not the fact, then, inasmuch
as the light of different wave lengths is different in
color, it is evident that one of the satellites of Jupiter, for example, on reappearing from behind the
planet after an eclipse, would be seen first of the
color which traveled fastest, the other colors subsequently appearing. Nothing of this sort being
perceptible, the inference is that in space all waves
have the same speed, whatever their length. Thus
the light that left far away suns, even thousands of
years ago, as ultra red, the slowest of ether vibrations that the eye can distinguish, has all that time
been traveling with a vibration frequency of 370
million millions per second. Likewise the violet,
the most rapid that can be seen, has kept all the
time its pace of 739 million million vibrations per
second with which it now meets our vision.

Is there no point, then, in all infinity that the
waves of light reach, travel worn? Without slack-
ening their pace, do they move on and on forever?
Through the ether, yes, as far as we know. But
when entangled with matter there sometimes comes
a change, a slowing, but not a progressive slowing.
Thus the ultra violet, which moves through the ether
of space with a vibration frequency of 833 million
millions per second, if caused to enter a solution of
quinine, will be given out blue or violet with a less

wave frequency. Or it may, on passing through other substances, be given out in several different wave frequencies ; or still again, taken up by decaying wood or otherwise becoming phosphorescent, it may give out all the wave frequencies that affect vision ; that is, it will have been changed into white light.

Therefore, although the view that some change comes over the waves of light on their journey through space is very weakly supported, even by inference, analogy still furnishes it a modified support. And just as we found that sound gives some indication of the distance it has come by a means which science has not clearly made out, so may light come from its far-away source with some quality that enables our judgment, acting all unconsciously, to recognize the fact — to form a dim conception of the length of its journey.

For even accepting as final truth all that has been adduced, the proof would not be conclusive that the vibrations of light do not become altered as they travel through space, only that all vibrations become altered in equal degree. It can hardly be that the ether is absolutely elastic, which it must be if each kind of light goes on the same forever. It is more than possible, then, that the waves of light do increase in amplitude and diminish in intensity with distance, and that violet light might somewhere in its journeying through the universe become red or some of the other colors of wave length longer than its own.

But even if this be not the case, even if the pace of light through the pure ether is forever the same, it does no violence to science to assume that, when light has become the energy stored in fruit and leaf, it takes on a character akin to the water wave and the sound effect into which it can be transformed. This admission does not, however, preclude us from seeking to discover that in some way the waves of light are affected as they journey through space, and that if we had a perception as much finer for light as light is more subtle than sound, we might still perceive this altered action.

This may even now be accomplished, either by refined impression on the subconscious intelligence or by its acting in a coarser way, that is, while in a process of transformation into other forms of force. For later we shall have occasion to inquire whether mind is not influenced by the forces given out in the metamorphoses of nutrition, and whether food in its retrograde changes does not give out in kind the forces that were employed in building it up.

MEMORY AND THE BRAIN.

Having thus briefly set forth the nature and the extent of the involvement of vibration in all manifestations of the common force, we may now proceed to consider the nature of mental or brain activities, with a view to ascertaining what analogies exist between them and the force-manifestations of external nature.

Memory has been authoritatively defined as the faculty of the mind by which it retains the knowledge of previous thoughts or events. The faculty may manifest itself in at least three different forms :

first, in the persistence in consciousness of impressions made upon the mind or by spontaneous recur-
rence of these impressions in consciousness, when it
is called remembrance ; secondly, in the recall of
past impressions by distinct effort, when it is called
recollection ; thirdly, in a form intermediate between
these two, by a conscious process of recalling past
occurrences, but without full and varied reference
to particular things, when it is called reminiscence.

In this essay, however, it is not proposed to
restrict the term memory merely to the retaining or
recalling of previous thoughts and other products of
brain activity, but also to embrace the method or
mechanism of their retention and reproduction or
reappearance in consciousness. In the fullest sense
memory may be regarded as the persistence in the
brain cells of all the accumulations of sensations and
perceptions in the form of ideas, emotions, and the
like, together with their susceptibility of being
brought up from time to time in such a way as to
be recognized in consciousness.

But before entering upon a consideration of such
of the functions of the brain as we shall be here
concerned with, it is indispensable to any satisfactory
understanding of the subject that we study, at least
in their relations, the different parts entering into the
structure of the brain that are concerned in mentation.

STRUCTURE OF THE BRAIN.

The brain, or that part of the cerebro-spinal nervous
system immediately concerned in conscious intellectual activity, consists of three divisions independent

one of another, and yet very intimately bound
together. These divisions are the cerebrum, the
cerebellum, and the medulla. Of these, the cerebrum alone is the seat of conscious thought-activity
or mentation, and for the purposes of this inquiry the
others need not be included.

The cerebrum or fore-brain consists of two lobes
or hemispheres, the one almost exactly and in every
respect the counterpart of the other, connected
with one another by a great number of white fibers.
These white fibers connect every part in each hemisphere with the like part in the other hemisphere, so
that the two lobes constitute a veritable twin system.

Each cerebral lobe consists of masses of gray
matter and closely laid bundles of white connecting
fibers. The masses of gray matter are composed
of vast numbers of cells arranged on the outer surface of the brain in the form of a thin convoluted
layer, which constitutes the cortex, and of other
cells in the central position, arranged in the form of
two gray masses or ganglions on each side, which
are coupled together and form the gray substance of
the optic thalamus and corpus striatum.

The white substance, which consists mainly of
nerve tubules, fills the space between the outer gray
mass and the central ganglions. These fibers partly
radiate like the spokes of a wheel from the gray
matter of the central ganglions to that of the brain
surface, and partly pass across from one hemisphere
to the other, connecting like parts of the two halves
of the brain ; and partly pass out to the general

bodily system, to carry motor impulses or to bring back sensations. These white fibers are constructed almost exactly on the plan of insulated telegraph wires or cables. There is first a central core of albuminoid material, which conducts nerve force and corresponds to the copper wire. Over this is a layer of fatty material, which is a non-conductor of nervous force, and corresponds to the non-conducting sheathing of the electric wire or cable ; and still outside of this is a protective sheath of connective tissue answering to the outer protecting sheath of the cable.

The gray matter is made up mostly of cells held together by a network of connective tissue. These cells are small at and near the surface of the brain, getting larger toward the central parts of it. They also differ much in shape in different situations.

The larger cells, which have the most irregular shape, have prolongations proceeding from .them which connect and become continuous with the central axis cylinder of the white nerve filaments.

Of these connecting prolongations, some of the larger cells have as many as seven or eight proceeding from them, some have but one or two, while many of the smaller ones have none, though curiously enough on occasion they will project elongations of their own bodies to form connecting filaments, as a leukocyte or white blood cell might do.

The optic thalamus is an ovoid mass of gray matter situated almost exactly in the center of the brain. It is made up of a series of ganglions ranged one behind the other, and two slender bands of gray

material extending down to and being continuous
with the gray matter of the spinal cord. These
ganglions of the optic thalamus receive the connecting nerve filaments from all directions, both from
the cerebrum and the cerebellum, and the general
system, through the medulla. It is the clearing-
house, the switch-board, the distributing point of
the entire sensory nervous system.

In front of the optic thalamus and closely connected with it is the corpus striatum, made up of
gray matter largely and forming the gateway through
which must pass the nerve fibers that carry motor
impulses from the brain to the rest of the body.

FUNCTIONS OF CELLS AND CONDUCTING FIBERS.

As bearing on the investigation with which we are
directly concerned, it remains to point out in a brief
way the functions of the several parts of the brain
immediately connected with thought activity.

Of the functions of the cortical cells, or those on
the outer surface of the cerebral hemisphere, it is
known that some are engaged in elaborating motor
impulses, some in the production of an electric current, others nerve force, and still others thought.
Doubtless if we shall ever attain to a full knowledge
of all the diversified work that is done by the brain
cells, it will be found almost infinitely complex and
infinitely differentiated.

It is probable that the optic thalamus is employed
in receiving sensations from all parts of the body,
sorting them out and distributing them to the various
cells of the cortical layer to which they appertain.

The office of the optic thalamus, as indicated already when speaking of its anatomy, was fairly then described as in most respects comparable to a switch-board in a telephone exchange. To complete the illustration, however, the aid of the corpus striatum should also be invoked, for it is to the motor vibrations largely what the optic thalamus is to the sensory.

The corpus striatum has for its office the sending out to the various muscles the motor impulses that are received by it from the motor cells in the cortex, an office relatively of the same character but apparently much less complex than that of the optic thalamus. We have thus in a very imperfect manner mapped out and described that part of the brain which constitutes the field of mental operations as a tangible stage upon which the imagination may watch the play of the wonderful actors in the drama of thought.

Hitherto no connection has been made out between the extremities of the motor nerves and the nerves of nutrition, which carry the outward current, and the sensory nerves, which carry the nerve current inwardly. It is certain, however, that an electric current is always going out on the motor nerves and returning on the nerves of sensation, and it is more than probable that as long as life lasts the circuit is completed by the tissues, and that a current of nerve force is continuously making the round of every nerve circuit, whether within the brain or in the general system.

By means of this arrangement an impression upon or a discharge of nerve force from any one cell may almost immediately extend to and affect many others, since nearly every cell in the gray matter of the brain and cord is connected either directly or indirectly with all the others. This description of the brain and its functions, though necessarily brief and defective, may yet serve to give a character of order and cohesion to the reflections that are to follow.

Utilizing all the principles set forth in the preliminary discussion in so far as they may be made available, we may now proceed to ascertain, if possible, in what way and in what form mental impressions are preserved in the cells of the brain, and in what manner they are from time to time brought into consciousness. We will try to ascertain what appearance memory or the contents of mind has in its secret home, even though the best solution we may hope to attain may still be regarded as largely hypothetical if not fanciful.

MECHANICAL THEORIES OF MEMORY.

Before attempting an explanation of organic memory, we must note certain phenomena that have been compared to it by way of illustration. Authors have found analogies for memory in the organic world, and particularly in that property possessed by light vibrations, whereby they may be stored up on sheets of paper or other material and there preserved for longer or shorter periods in the form of latent vibrations, to reappear at the summons of a developing agent.

Engravings exposed to the sun's rays and then kept in a dark place can months afterward, by the aid of appropriate reagents, be made to yield persistent traces of the photographic action of the sun upon their surface. Again, if a key or other like object be laid upon a sheet of white paper, and the two be exposed to the direct rays of the sun, and then the paper be laid away in a drawer, years afterward the spectral image of the key will still be visible.

In these examples we have to do with but a single condition or element of memory, namely, the retention of the impressions. Furthermore, this retention of impressions is only in a passive form. The incessant changes, the endless disappearances and reappearances of the contents of memory, have here no parallel. If one's ideas and other mental furniture are simply photographs, merely dead images, how are they to be awakened into renewed activity? Above all, how are they spontaneously to awaken one another? We may go on hour after hour with a vivid train of thought, with all our senses completely closed to excitations from without, thoughts, ideas, and emotions in an endless train arousing one another. Mere dead pictures could not act in this way. Photographs may be heaped mountain high, but they could no more awaken each other than could the unused plates upon which they have been impressed. Even though it may be that the dead can bury their dead, we may rest assured that the dead can never awaken their dead.

MEMORY AN ACTIVE PROCESS.

That which produces images in the mind, perpetuates in an active form the contents of the mind, must therefore consist of elements constantly in motion — of virtually living essences.

It might be assumed that the intercellular forces that have been described have something of the character of the cable car, where the cable is constantly in motion and the load of the car added to from time to time at proper stations ; or still more like transmission by telephone, where the current is continuously active, but is modified into intelligible manifestations of force by the vibrations of the diaphram. The currents of force among the brain cells are supposed never to cease as long as the life of the individual lasts.

It is, however, only a crude parallel to the operation of mental forces that we may find among mechanical contrivances. We will endeavor, then, to find what relation the elements of memory in their behavior bear to the common forms of force as exhibited in undulatory motion. It has already been shown that we know nothing of force except as a mode of motion, and that motion in all cases appears to take the form, potentially at least, of vibrations or undulations ; and that since it is a necessity of thought that action and reaction are equal, it is impossible to separate the notion of undulatory movement from the action of force.

We have also found in all our investigations of transmission of force by means of undulations some reason to believe that such transmission is characterized by increase in amplitude and diminution in intensity of vibration. In this light, then, let us examine the phenomena of memory and ascertain if they do not present analogies sufficiently strong to make it highly probable that they also depend upon some form of force expressed in terms of undulations or waves and correlated with other forms of force in nature.

I would have it understood that I propose to make no attempt to explain the nature of consciousness. Consciousness is accepted as something totally incomprehensible and inexplicable in any scientific sense. Nor can any new facts or new observations be offered here in the elucidation of memory. The aim is simply to apply analogies drawn from well-known facts, some of them remote enough indeed, in the hope that at least an additional glimmer of light may be let in upon a most obscure and difficult question, and that in this way some additional aid may be given to whomsoever will pursue it to a satisfactory solution.

ANALOGIES OF EXPRESSION IN ANIMALS.

The first of these analogies may be found in the similarity of terms and movements by means of which nearly all classes of animals are accustomed as a rule to express similar feelings. Thus it has ever been to thoughtful men a question of curious and abiding interest how it results that each class of feelings among men and nearly all the higher animals is expressed in similar and seemingly corresponding

tones and movements. The voice of anger, the cry
of joy, the wail of grief, or the cooings of love we
may usually recognize, even though we may never
before have met with an individual of the species
from which they proceed. Whatever the source,
the utterance is perhaps invariably the same. These
different tones and expressive movements, the true
language of their feelings, have not been learned by
the different species of animals from one another, and
it is also certain that they have not been derived from
any remote common ancestor by inheritance. For
they never could have been possessed by any possible common ancestor except in the most rudimentary
form. Yet beyond all reasonable doubt the employment of these tones has originated from a common
source ; all animals have derived them from a store of
influence that operates upon every living thing alike.

What is that source and what are the laws by
which these results have been obtained ? For,
whatever views may be held of the origin of life, we
can not doubt that all things are effected now not by
miracle but in accordance with the laws that infinite force manifests in nature. The point to be made
is, that all mental operations this side of consciousness are due to vibrations or undulations, which are
the medium of expression of the forces that are
found active in the part of the brain which is the
seat of consciousness ; and that the impact of these
undulations, whenever it is sufficiently strong, whenever the strokes are of sufficient intensity, results in
an act of recognition by consciousness. The persistence of these waves or undulations in the brain
cells is what we rightly call memory.

The first vibrations resulting from the striking of a tuning-fork or a bell may be taken to represent the various forms of impressions made upon the sensorium, such as light makes through the eye or sound through the ear.

The bell or the tuning-fork continues to vibrate, and in this we have the counterpart of memory; only in the brain the undulations continue indefinitely. Memory, then, is only a peep we gain of the mind at work.

ANALOGIES OF MIND AND COMMON FORCE.

We may now particularize and trace out some of the numerous and striking analogies between the different forms of the common force of nature and the affections to which the mind is subject in receiving, retaining, and reproducing impressions.

We have already seen that the undulations or waves in water and in the vibrations of earthquakes become slower and longer as they advance, and we have found, also, that there is some reason to believe the same to be true of light and sound.

Furthermore, we have good reason to believe that the same laws apply to all forms of force ; for since they are mutually convertible the one into the other, we can not well doubt that they are subject to the same laws of action.

RESEMBLANCES BETWEEN VIBRATIONS AND MEMORIES.

Turning now to the examination of memory with reference to these laws, it is found that ideas of

things far removed in time or space are called up
by suggestive impressions that reach us through the
medium of waves of large amplitude, that is, of long
and slow waves, while quick movements, suggestions
presented to us in terms of short waves, call up
ideas of things less distant, and also of those more
recently formed.   Thus the view of lofty mountains,
dim and azure hued in the distance, vistas opening
out into the ocean, and above all the strains of low
pitched, soft and plaintive music, fill the mind with
reminiscences of things that are far away, or of
events of the almost forgotten past.

And thus it is that for that instrument of music
that gives the sweetest, softest, and most melting
of tones, ''far windharp'' is everywhere the self-
suggested name.

On the other hand, quick and lively tones, spirited
music, necessarily made up of short undulations,
chain the thoughts to the present, and instead of
calling for stillness, silence, and reverie, prompt us
to responsive and suggestive expression in action, as
if the thing were in easy reach and easy of attainment.

No one springs at the sound of far-off cannonad-
ing, or the rumble of distant thunder, any more
than one feels moved to dance by strains of slow
and far-off music.   That which appears to be in
reach, that which is made sensible to us in terms of
quick vibrations, re-enforces the vibratory currents
in the brain that have to deal with quick motions,
and thus brings these motions definitely before consciousness.   On the contrary, that which is made

sensible to the mind in terms of slow, of ample vibrations, re-enforces those undulations in the brain that have become slow ; vibrations that are the vehicle that perpetuates in memory impressions of things that are long past or seemingly far away. And thus the body is impressed to stillness because the source and cause of these vibrations are instinctively recognized as beyond its reach. The slow undulations in the mind are not the ones that give rise to bodily activity. The body does not attempt to reach what is instinctively felt to be out of reach.

"Be still and know that I am God" was an injunction justly suggested by an all-pervading truth, in that it is in stillness alone that the mind may even attempt to grasp the infinite.

This relation of fitness between the mode of suggestion and the characteristics of the thing suggested, between the language employed and the thing it describes, has found universal expression in the choice of tones in music and speech, of measure in poetry, in the blending and arrangement of colors in painting, and, in short, in all the delineations by which art has sought to impress the mind through the portrayal of the truths of nature. "It is, indeed, the law, even the holy law, that imposes beauty on the artist."

In every land and in every age men seek to conduct the exercises of religion with a solemnity related to and suggestive of the objects of worship. If the object worshiped be regarded as limited in power, and imperfect in its divine attributes, the tones and other accessories employed in its worship are not greatly different from those employed by men in their intercourse with one another.

To the Greeks, Jupiter, who could sit no higher than the crest of Olympus, which they themselves could easily scale, was but a superior Greek. But when the divinity had ascribed to him the vast or infinite power of a Brahma, with his seat in the dim heights of the Himalaya, or a Jehovah who could thunder from Sinai, soft, deep monotones, slow and submissive, were chosen for the language of worship, far-away refrains for its music, and even subdued colors for the habiliments of its votaries.

The aim in religious exercises dedicated to objects of such might and power is to give expression to feelings and ideas awakened by the contemplation of a being vaguely conceived, vast, mighty, kindly, and far away ; therefore are chosen such music and tones of speech and other accessories and surroundings as produce sense waves in the organ of mind most nearly akin to those by which impressions of an object vague, vast, kindly, and far away would be communicated to it ; for undulations of this ample character are the elements out of which our ideas of such objects are formed. Hence the spacious cathedral, the deep-toned organ, and the echoing cavern as accessories of worship.

So the child will speak of an object perhaps but a short distance away, but which in the untraveled pathways of its brain may have impressed its mind as being at a great distance, as ''away off yonder,'' in slow and measured tones that plainly answer to an unconscious realization of the form in which such impressions reach the sensorium.

There are said to be savages who have in their language no change of terms for comparison, but who are accustomed to express different degrees of quality by differing tones of voice.

Thus the same word may mean a rill, a rivulet, a brook, or a river, according to the tone in which it is spoken. In view, then, of these facts and a vast array of similar ones that might be adduced, is it not reasonable to conclude that the elements of our ideas correspond in their nature to the undulations proceeding from the objects they represent, and that they are built up of the waves of light or sound or other sense-impressing forces that come from without?

THE THEORY PRESENTED.

Having proceeded thus far with the induction, it may now be well to state more fully the theory which the foregoing facts tend to elucidate. It assumes

First: That in the cells of the gray matter of the brain, and possibly of the spinal cord, orderly groupings of waves or vibrations are set up among certain atoms or molecules by waves of light or sound or whatever else may affect any of the senses.

Second: That these undulations are realized first as sensations; that they then group themselves in such a way as to form perceptions, ideas, emotions, and various other products of mentation.

Third: That these vibrations or groups of vibrations, passing from one cell to another along

the connecting white fibers, or from one part of a
cell to another part, rise in succession into the scope
of consciousness, and that these groupings occupy
the attention of consciousness in the order of the
number and intensity of the waves of which they are
composed.

Fourth : That by reason of some unapprehended
property, consciousness after a time either ceases to
be affected by any given force of vibration or else
the vibration itself is borne on by the constant current of force persistent in the cells, from the point
of contact with the seat of consciousness, and that
when this occurs, other vibrations may enlist the
attention of consciousness and gain its recognition.

Fifth : That there pass continuously through
and among the cells, normally, innumerable vibrations or groups of vibrations, the residua of former
impressions, which are too weak ordinarily to force
themselves into the scope of consciousness, but
which yet may gain its recognition by taking to
themselves other waves of the same character as
themselves, gathered from those already present in
the brain cells or else derived immediately from
without.

Sixth : That the tendency of all these waves
and groups of waves, like the waves of light or sound
in space, is to grow gradually slower and weaker,
until in all their combinations, if left to themselves,
they would escape the utmost grasp of consciousness ; but that when re-enforced by kindred waves
they may again enter into the scope of consciousness.

Seventh : Finally, the theory assumes that sensations, perceptions, ideas, emotions, and all similar outcome of mental action or mental experience whatever are essentially one and have for a basis the same ultimate elements, namely, sense waves or undulations combined with the normal force movement residual in the brain and variously modified. Memory is supposed to be nothing else than the persistence of these undulations in the nerve centers, with or without the power of distinct reappearance in consciousness.

As an example and probably an instance of the reinvigoration and revival of vibrations that have become latent in the brain, let us suppose that a sense wave derived from any of the senses, say the sight or hearing, enters the visual or auditory center of the brain and there finds a group of homogeneous waves that have been long present but become too weak to command recognition from consciousness. The result will be that the ideas which these groups of waves once constituted will be again revived and brought into consciousness.

When this occurs the predominance in potency may be either with the sets of undulations already present in the cells, or with the newly added vibrations, and in proportion as one class is stronger than the other will that class be recognized by consciousness as mainly constituting the new idea. If the sets of undulations already present in the mind predominate, the newly brought waves will not in that case be distinguished or separately recognized by consciousness, and the product of the combination will be regarded as an act of reminiscence or recollection, sensibly or insensibly modified.

But if, on the other hand, the newly added vibrations preponderate, a seemingly new idea or mental affection will be formed, which will, however, be more or less dimly colored or tinted by the fading undulations already become latent in the brain cells.

Under such circumstances the vanishing undulations so recalled may arouse vague feelings or impressions of the far beyond, or of the long forgotten past, in connection with the new idea.

Through the dim shadows of mental images thus formed there will often flicker glimpses as of life and scenes in another world or another state of existence. As Richter says of music, they may lead us to think of things we have not seen and yet shall not see.

A further analogy, drawn from a practical illustration of the laws of sound, will serve in some measure to make plainer how vibrations from without, of the same nature as those supposed to exist within the brain as the remnant of former impressions, can enable these to arouse consciousness and manifest themselves as memories.

It is a law of acoustics that the multiplication of a sound does not increase its intensity ; that is, a sound produced by many voices will proceed no farther than when produced by one. Yet the sound of many voices together can be heard farther than the sound produced by one. A single bee may be buzzing near us and not be heard ; but if another or several others join it in the same situation, together they may produce a distinctly audible noise. It is thus with the undulations that in the ultimate home of consciousness have become the mind's record of experience, the memory of things.

With the lapse of time they have become less and less vivid, less and less intense, less and less distinct, until they have settled down, where they bid fair to remain beyond the utmost reach of consciousness. But the sight or hearing or taste, or some other less specialized sense, brings from without a kindred wave, or within the cells new couples are formed among the participants in the intricate mazes of the never-ending dance, or even many pairs and groups may join in noisy promenade. So strengthened, they force the ears of consciousness to hear, and thus ideas and emotions — though never the same as before but always modified — are made to appear on the open arena of mental activities. It is no more possible to think the same thought twice, to feel the same emotion twice, than to see the same cloud twice journeying through the air, or a smoke curl twice floating in identical form.

GROWTH OF IDEAS.

Consistent with the views here advanced are our observations of the growth of ideas, or of sensations into ideas. It will hardly be doubted that sense vibrations are recorded in all minds in a similar manner, provided all the senses of the individuals are of the same nature, that is, normal; therefore, all the obvious attributes of an object will produce a similar impression on every mind. It would seem that expression should also be the same, and that word pictures of ideas or what is known as onomatopœia should be the rule.   But this is far from being the case.

How is it, then, that direct designations, such as the names of objects, although never so multifarious and having apparently no suggestive relation, will call up the same mental image in different minds?

For instance, take the name of one of our domestic animals ; say the horse.   Its name in any of the various languages, such as hippos, equus, pferd, caballo, or cheval, will to any one understanding these languages suggest an individual image.   Yet not one of these names is a thought word.

Can it be that there is in nature one true language, and that in the nerve centers all other languages are translated into that expressive tongue?

It is much more likely that the undulations which in the mind form the sensible image of an object enter into groupings with the undulations that form every other attribute or name that such object is perceived to acquire, and that its idea grows larger with every new name and attribute.

Thus, though we designate an object such as a horse, for instance, by a hundred different names, each of such names, whether heard or seen or made out through the touch, as with the blind, will form part of the complex idea of a horse, just as will the hoof or head or mane or tail ; and sensory undulations proceeding from any one of these will be sufficient to call up in consciousness the idea of a horse.

Every attribute becomes a handle by which the

idea in all its complexity may be summoned into consciousness.

It is easy, then, to account for the fact that men seek factitious elements of greatness in the way of complimentary mention, imposing dress, ornaments, and decorations.

The tattooed savage and the thrice crowned and bejeweled potentate are traveling in the same path.

In harmony with the foregoing theory is also the fact that the tenacity with which the mind holds impressions made upon it bears a direct relation, other things being equal, to the length of time the vibrations from a given source may have continued to play upon consciousness. A like result is also accomplished by intensity of impact.

Thus, after having suffered from certain forms of infectious fever, such as acute rheumatism, for instance, men have been known to have let slip from memory all knowledge of the events of a short period previous to the attack of illness. Similarily those who have suffered from severe concussion of the brain, or a stroke of apoplexy, will often be found to have forgotten impressions made upon the mind at a date just anterior to such accident, unless such impressions shall have been made in a very vivid manner.

The undulations in these cases, in their circuits among the cells, not having had the time or initial force to enable them to enter into a sufficient number of groupings with others in the brain, not having had time to form a sufficient number of acquaintances, so to speak, found no friendly hand after the shock to awaken them, strengthen them, and aid them in demanding recognition at the hands of consciousness.

APPARENT SHORTNESS OF TIME.

In a similar way, by a converse process, we may account for the seeming shortness embraced in the period of our past experience. All the long years of life that is past seem but as yesterday. "Time flies" is a maxim current everywhere.

The reason of the seeming shortness of time appears to be that the undulations which in the brain constitute the record, the memory of each hour, are molded into and kept alive by those of the hours that follow ; and thus closely united and blended together in the memory, years become days and weeks become moments.

We call time short, in part at least, because the past is made to seem so by the mechanism of memory.

So, also, those who are associated with us from childhood or early life never grow old.

The brother or sister, though silver-crowned and bowed with age, is the brother or the sister of the play place still, and venerable companions of wedded life, "tottering down the hill together," are still, the one to the other, the manly groom and the radiant bride with the orange blossoms at the altar.

DREAMS LAG BEHIND EXPERIENCES.

A further illustration may be drawn from certain peculiarities of dreams though the phenomena here presented might be susceptible of other explanation. It is a peculiar feature of dreams that changed circumstances are seldom realized in them until a considerable period of time has elapsed after the change.

The scenes that are presented to us in dreams, after we have made change of residence, for example, are for a longer or shorter period of time almost invariably laid at the place of former residence and surrounded by familiar features. This may be accounted for, in part, upon the supposition that nearly all motions, both about and within us, while we sleep, are as a rule slower than those of waking hours. The breathing is slower, the heart-beat is slower, and, besides the general stillness that environs us, the senses are closed to excitations from without, so that the memory waves of the past, undisturbed by fresh undulations, have freer play upon the half-awakened consciousness.

Night, even without sleep, brings far-away memories, and few indeed there are who may not with truth repeat with the poet :

> "Oft in the stilly night,
> Ere slumber's chain has bound me,
> Fond memory brings the light
> Of other days around me."

No one can realize in dreams till after some lapse of time the fact that a great calamity has been experienced in the hours of waking. Bitterly as the waking hours may be haunted by the memory of some cruel loss, some stinging sin or searching sorrow, we may still lie down to pleasant dreams, and then awaking to a troubled appreciation of the painful reality, we fain would persuade ourselves that the truth is itself the dream.

"THE RAVEN."

Nowhere does this principle seem to me to have received a more appropriate portrayal than in that most weird of poetic productions, "The Raven."

The half-revealed theme of "The Raven" seems to be our experience of the method in which painful happenings of waking hours take possession of our dreams, though the original inspiration may have been in some measure lost sight of and obscured by the poet in bending the description to the requirements of artistic rules. Well chosen is the poet's vision to symbolize a biting sin or heart-searching sorrow, with raven aspect, stalking in dreams into the halls of memory, thrusting its beak into the heart, and fated by and by to gather all the soul into its never-lifting shadow.

When we consider the number of undulations necessary to record all the sensations and sensation products of a lifetime, the magnitude of the number, as indeed in every theory of memory, seems to offer an insuperable difficulty. This will disappear, however, when the possible extent and rapidity of undulating motion is taken into consideration. In the transmission of light the waves of the conducting ether are said to obtain the rapidity of four hundred and twenty eight million millions of shocks or waves per second for red light, to seven hundred and thirty nine million millions for violet, and a number vastly beyond these for the transmission of chemical or ultra violet rays. But inconceivably great as these numbers may seem, they form only an infinitesimal part of the undulations that can take place in a single direction out of the infinite number of directions possible at the same moment of time. Let us suppose the visible universe to consist of a hollow sphere, and every one of its hundred millions of sparkling stars to be a luminous eye giving out light, and at the same time gazing at its fellow on the opposite side of the sphere and directly through the central atom.

For a ray of violet light, that central atom must dance in waves to the number of seven hundred and thirty-nine million millions per second for each one, or at least for each pair, of these starry eyes.

And this is in but one direction and for one variety of light.

Yet between these two stars, and still in the same direction, this atom must further keep step to the vibrations of at least eleven other varieties of light, or twelve, if the ultra violet be included, ranging from three hundred and seventy million millions to eight hundred and eighty-three million millions per second.

Now take two other of the sparkling eyes half round the vault from these, and looking at each other directly through this same central atom. This atom must now, while maintaining its former motion, vibrate in a directly transverse way with the same rapidity as before.

And after all, this is only a beginning; for these vibrations must all be repeated fifty if not a hundred million times 'for the entire host, and at the same moment of time.

And yet still multiplied millions, as many of these eyes might bestud this limitless vault and each see its opposite through the atom of ether at its very center. Similarly, then, at the seat of consciousness, each atom, and much more, each complex molecule, may, by its infinite possible undulations, be employed at one and the same time in a like infinite number of sensations, thoughts, ideas, emotions, and other contents of memory.

It is well known that the apparatus of mind is not responsive to all modifications of force that are operative in nature, and that some that are operative are not consciously perceived.

Our sense of hearing does not perceive undulations of greater frequency than forty-two or forty-three thousand per second. The rate and range of the undulations perceived by taste and smell have not been ascertained, but the presumption is that they are vastly more limited than those made sensible to sight. It has been ascertained that sight is insensible to undulations of less than three hundred and seventy million millions or more than seven hundred and thirty-nine million millions per second, and it may be that even within these limits there are vast numbers of undulations that make no conscious impression on the senses, vast numbers of moving fingers that find no keys in the organ of mind upon which to play.

Outside of these limits are doubtless untold numbers that by our senses alone we may never apprehend.

OF THE NATURE OF EMOTIONS.

This brings us to a point in the discussion where a more extensive explanation is required of certain terms already used, in order to intelligently consider the nature of such experiences as grief, joy, and other similar affections known as emotions.

It has already been set forth in the statement of the vibratory theory of mind that sensations, perceptions, ideas, and emotions, with all other outcome of mental action, are essentially one, and that they depend for their difference upon the different groupings of vibrations that play upon the seat of consciousness and the different direction and intensity of the play. This would assume that an idea is not a visual image but a grouping of vibrations as they play on the seat of consciousness, a musical image rather than a visual image.

We have seen how undulations that enter the brain as sensations might build up ideas into very complex forms. But before that point is reached many distinct changes must have taken place in the groupings of the undulations.

A mere sensation to the eye of consciousness has no shape ; it is simply the ringing of the door-bell. It may be anybody's ring. But when it begins to take form it becomes a perception, and, strictly speaking, should no longer be called a sensation. We speak of a sensation of heat, and custom justifies it ; but ought it not be called a perception of heat ?

So with color, we even say sensation of color, after we have taken the color into definite comparison and it has become a mental image, an idea.

Now, when any thing has produced an image on the organ of the mind, whether a visual image or a grouping based on the other senses that has reached the same stage, we call it properly an idea. An idea, then, may be defined to be such a grouping of sense waves as are capable of producing in consciousness images or suggestive representations of external things, that is, external to consciousness.

An emotion is simply an idea which has had woven into its complex group a greater or less number of such undulations as are otherwise usually employed in the production of any kind of expressive movement, whether such expression be brought about by muscular movement or secreting gland. In short, it is an idea plus any kind of involuntary movement due to it ; hence the term emotion, or moving out. This employment of the term, however, is not without exception, if it be not itself the exception. The term is often applied to the physical expression of the feeling, and in that case the emotion would not embrace the idea, but would be merely a kind of muscular effort produced by a certain class of ideas. As long as these complexes of undulations accord with the well-being of the individual, the result is happiness ; when it fails to do so, the result is unhappiness — the idea or emotion becomes disagreeable.

GRIEF ANALYZED.

The analysis of one of these emotions will serve to supply a better understanding of what is meant, and this analysis is perhaps best attained by the study of the evolution of an emotion. Suppose I have any commonplace object that is of an agreeable pattern, a painting, for instance, and that the undulations which make up its mental image meet with an agreeable response in consciousness. I lose it, and at once feel that there exists an unpleasant vacancy in the assembly of mental impressions. Something, for a short time at least, is lacking from accustomed pleasant influences.

The painting can no longer send the required undulations to re-enforce its agreeable mental image, and I have therefore a disagreeable mental experience, namely, that of a blemished, maimed, or otherwise impaired idea. A myriad of similar experiences, very many of them connected together by being made up of undulations common to all associated cells, enter into the make-up of my state of mind. So far we have assumed that none of the undulations going to form the idea take hold on or belong to motor cells in the brain, and therefore elicit no stimulus to involuntary expressive movement. But if repeated every day, or often, the loss idea might extend to the motor cells and take on the character of an emotion. It may be, however, that great numbers of small or more or less vexatious losses have been suffered. I may then reach a point where the loss of my painting, in addition, may cause a very painful idea. And why? For the simple reason that the memory record of these successive marred ideas has been kept in the same or a closely connected area of the brain, and with each new loss they are called up, and combine to make a more painful impression.

The case is similar with emotions. We have a loved one, say a mother, whose idea has been enlaced with our being by innumerable agreeable undulations until it has taken root in the sources of expressive movement. A thrill comes unbidden even when the name is spoken. We lose her by death. We, of course, have never lost a mother before, yet our hearts are filled with the deepest grief. Now this grief is produced only indirectly by our mother's death; for how could we know to grieve so bitterly for that we never before experienced? The direct and potential cause is the marred and wounded emotions left in the mind by all the past partings and bereavements of life, now summoned up by this fresh wound, whose record is added to their own.

The proof that such is the case is found in the fact that very nearly the same effect will be produced by a false report of our mother's death, such a report being sufficient to open the sluice-gates of marred and withered emotions.

INITIATION OF REMEMBRANCES.

But who knows how or what is the first stir that marks the beginning of this movement? It may be that some vibration joins a related one somewhere in the brain cells, and the two start up a diminutive electric current that sets the machinery in motion, or stimulates the nerves that dilate the arteries and bids them allow a greater current of blood to flow to the cells where these memories are stored, in order that they may spring like a flash into the arena of consciousness. We have seen that close beneath the surface of the brain there are large numbers of cells that ordinarily have no filaments or nerve tubules connecting them with the other cells, but that occasionally, when they need to communicate with their fellows, they stretch out portions of their own bodies — string their own wires, so to speak — and establish communication with brother cells of the same class, and also with the general mass of brain cells.

Vibrations of some kind are doubtless stored in these isolated cells, and when in their tiny halls maneuverings take place, they feel prompted to report the result, or it may be they desire to get fuller reports of what they hear passing on the wires, or otherwise reaching them as emanations. Nor is there any thing more strange in the fact that conscious impressions may be made upon brain cells, apart from those that come through the senses, than that these tiny independents should learn what is going on about them and open up communication with their fellows. The phenomena of sympathy and of suggestion, which is closely related to it, also find a possible explanation in the principles here presented, as will be later shown.

CONSCIOUS AND UNCONSCIOUS BRAIN WORK.

The part of brain work that is consciously performed is almost as nothing when compared with the unlimited work that goes on unconsciously. To a thoroughly predominant extent the *timbre* of life, the *wohl-ge-fuel* of the Germans, depends upon the character of unconscious processes. We will assume, then, that muscular movements of every kind are a part of thought, an attachment, or possibly better, a metamorphosis of thought, either conscious or subconscious. Certain groupings of vibrations, taking appropriate direction, determine a given set of muscular movements. Even an emotion we have found to be an idea which has been so heavily laden with vibrations that a part of them escape in the form of physical or muscular movements; or, if better expressed, a muscular or a secreting-cell movement involuntarily accompanying an idea.

The movements so produced are in all probability the correlates or multiples of the undulations that caused them. If this be true, we can form a dim conception of how the perception of a particular movement will be followed in the observer by a retranslation of that movement into the undulations of which it is composed.

Say, for instance, I observe some one to yawn. Assuming that the act of yawning is decomposable into the undulations which, in the brain, give rise to the act ; that these reach my brain through my sense of sight, and there arouse and intensify the corresponding undulations with which my brain produces the act of yawning, a muscular act will follow identical with that which I have observed in another, accompanied also by the appropriate feeling. The case is the same with laughter, weeping, and all other outcome of emotion. If we assume that tones of voice are intertranslatable with the nerve vibrations that determine their production, we can easily see how angry words bring an angry response, and how "a soft answer turneth away wrath," and even why it is that "all the world loves a lover."

So it is that a deeper interpretation may be given to the all too much neglected observations of the gifted Delsarte on voice as indicative of character. If one man, as he shows, expresses himself with harsh, guttural tones, and thus indicates harshness and coarseness of character, it is because the vibrations in the brain cells, that are the primary origin, and the constituent elements from which such expression is formed, are themselves coarse and harsh, and, therefore, all the actions of such a man will show a logical consistency, as indeed will the structure of his physical organism. On the other hand, the man or the woman who employs soft, palatal tones, whose words seem to linger with a sweet taste in the speaker's mouth, speaks also from the gentle vibrations that store the experiences of life in the brain and gives evidence of inspiration from a source marked by greater refinement. Indeed, it is not impossible that nerves of nutrition have their burden of vibration influenced in a similar way. It has often been observed that after years of association husbands and their wives come to resemble each other.   Likewise it has been frequently remarked that children born of foreign parents, or going to foreign lands, come to resemble those with whom they associate in a new country in a way that can hardly be accounted for by the influence of climate. However, we shall see further on that an emanation more subtle than that with which the senses are concerned may also determine the influence of one mind upon another, and the physical movements resulting therefrom as well.

MIND-READING OR TELEPATHY.

Among the mysterious phenomena of mind that seem to some extent susceptible of explanation on the theory here advocated is that of mind-reading.   It is not a violent surmise that there exists a world of gentle, subtle undulations operative in the organ of mind that yet, with most of us, never come distinctly within the purview of consciousness.

That objects of various kinds give off delicate vibrations which those who are blessed with sight never perceive is indicated by the fact that people absolutely blind have been enabled by the sense of touch alone to accomplish the most delicate distinction of colors.   This doubtless is effected by the recognition of undulations that must be present with us all but only very exceptionally perceived.

How reasonable this is, an observation made by every one who has watched the change of day into night on railroad trains bears ready evidence.   During the day the glass in the car windows seems to offer no reflection of any of the occupants of the car. But as soon as darkness settles down, each pane of glass becomes a mirror and presents a vivid reflection of the contents of the car.

This is not because a larger number of rays of light are being now reflected back than during the day; on the contrary, there are far less, for the rays were more abundant during the day. But during the day consciousness was dazzled and blinded by the flood of vibrations that came from without and did not recognize the small number reflected from the glass.

Few individuals have realized how much pain can be inflicted merely by the operation of the healthy process of nutrition — the tearing away by the leukocytes of the used-up cells — in our bodies until they witness the sufferings of those who have for a time blocked the pathway to consciousness by the habitual use of morphine and then quit the drug. The whole body is then filled with pain most difficult to bear, and yet the same activities, the same causes of pain, though in full operation, are in normal conditions unobserved. Some individuals possess the power of colored audition, which can hardly be other than a peculiar power of analyzing sound into still finer undulations, viz., the ether vibrations that are the factors of the waves of sound. Under the influence of hydrophobia and some other affections the sensibility of the deaf has been known to become so exalted that they could hear acutely. There are other vibrations still that we know must fill all visible space which have never yet been brought

directly through the medium of the senses into
human consciousness.

Thus tremors apparently produced among stars
so far away that they seem to border the universe
wholly unperceived by human sense have been
demonstrated through unstrumental measurement.

In view, then, of the foregoing and many other
similar facts, may we not with reason believe that
the mind is all the time receiving from the external
world countless subtle influences, subtle waves of
the nature of those that produce sensations, but
ordinarily not of sufficient intensity or the proper
pitch to be recognized?

These influences or *quasi* sensations ordinarily
form only the coloring of thought, perhaps only the
vaguest of feelings, even the drapery of emotions ;
nevertheless, in brains peculiarly sensitive they may
become in some degree manifest as the material of
thought. If, then, we add to our theory the further
hypothesis that there is a delicate sense possessed
by men in common with other animals, and as yet
unnamed, and admit that there exists in nature the
obscure forms of force referred to capable of affecting that sense, in some organizations even to such
extent as to secure definite recognition at the hands
of consciousness, we should have supplied to us an
explanation of mind-reading and other kindred
powers that many individuals are known to possess.

The question of mind-reading has met with much
skepticism, and doubtless far too much has been
claimed for it. Yet no fact in science has been
better established in so far as it is necessary to con-
firm the principle.

The transfer of the motor impulses from mind to mind without any medium of sensation seems to be the form most commonly observed. Thus the so-called mind-reader may discover a hidden object under circumstances calculated in the highest degree to baffle and mislead; will sing a song silently thought over by some one near; or perfectly blind-folded will drive a team at a rapid pace through a crowded street, guarding against a collision as effect-ually as the most expert driver with all his faculties free. In all this he will ask only to be kept in con-tact with his prompter, and in the driving feats this is simply the touching of their feet or knees together.

In view of what has already been set forth, this need not seem at all strange. If we can suppose the groups of undulations in the brain of the prompter to have such force as to extend to that of the recipient and agent, and then to be discharged to the muscles of the mind-reader, just as they would have been discharged into the muscles of his prompter, there is nothing more inexplicable in it than there is in the fact that one's own muscles are obedient to the brain that directs them.

EXPERIENCE OF MOLLIE FANCHER.

There are, however, phenomena of this class still more remarkable, still harder to believe, and yet far better authenticated than the facts in any system of theology in existence. In this category may be included the well-known case of Miss Mollie Fancher, of Brooklyn. One particular instance in her history is especially in point. A number of most trustworthy

gentlemen having clipped into small pieces a leaf of
a book, which in order to prevent impressions being
made on their minds they were careful not to read
nor even to see, they by turns put the pieces into
three envelopes, one within the other, and submitted
the package to Miss Fancher.

She passed it over her forehead, and then, with
her hands behind her head, wrote out the contents.
Every now and then she would make a mark for a
blank and then proceed with the writing.

What was the surprise of the investigators to
find, on reaching home and placing the pieces
together in their right order, that the dashes corre-
sponded with certain of the clippings that they had
unwittingly let fall and left at home on the floor, and
that otherwise the original had been accurately
reproduced! True enough this was not mind-reading, but it serves to illustrate the action of insensible
waves on a most delicately organized brain.

We have here to suppose that delicate and subtle
undulations, probably related to fluorescent or phosphorescent rays, were given off from these clippings
and reached Miss Fancher's brain somewhat in the
way that colors make their impress on the brains of
the blind. We must further assume a brain so sensitive that it could focus the bits of paper as the eye
does visible objects, or measure the distance by dif-
ferences in the vibratory force of the emanations,
and being so retained in mental vision they could be
sorted out and arranged in a manner similar to that
pursued by the experimenters themselves in making
proof of the translation.

We may safely say that there is no limit to insight of this character except the dullness and coarseness of our nervous organism. Nature everywhere around us must needs be vibrant with revelation if we were only so organized as to be able to receive it.

EMPHASIS AND INFLECTION.

When a tuning-fork, a string, or a pendulum is set to vibrating, the amplitude or extent of excursion of the vibrating body on the two sides of the middle point or the point of rest is the same. That is to say, as far as the vibrating body goes in one direction it will go also in the other. In another than the ordinary sense, action and reaction will be equal.

Close observation will show that this law operates to govern emphasis and inflection in speech and even the structure of sentences. Few things add more to the beauty of composition than the proper balance of clauses. If one should quote the maxim, "I was once young but now am old, yet have I never seen the righteous cast down," he would feel that it was incomplete, the introduction outweighing the conclusion, and a sense of relief would be brought by adding the clause, "nor his seed begging bread."

A question and answer together constitute a harmonic period; the voice goes up in asking the question and comes down equally in giving the answer. This may be observed even among lower animals. A lamb, a calf, or a colt will bleat or whicker with the rising inflection, as much as to say, "Where is

my mother?" and the mother will answer with the
falling inflection equally as plainly, "She is here."
Emphasis can be successfully subjected to analysis
with like results.

If one word in a sentence or the words expressing one idea is the subject of emphasis, some other
word or set of words must receive an equal amount
of emphasis in an opposite direction ; that is, in the
rhythmical excursion the positive and negative elongations must be equal.   An entire chapter, however,
would be required to fully set forth the principle in
the fullness of its application in this connection.   It
must suffice here to claim that emphasis and inflection are decomposable into the finest vibrations —
even the ether vibrations.

THE SOURCE OF MORAL LAW.

As a corollary to the doctrines herein hinted at,
when referring to the law of beauty it must follow
that all recognition of truth or the true standard of
the fitness of things is the result of an attempt of
the mind to harmonize impressions received by it
from without with the rule framed for its guidance
by means of the orderly undulations of ether, manifested in the transmission of radiant force.

Out of the various combinations of the rich store
of accumulated undulations coming to us originally
either directly or indirectly, in the shape of light or
heat or actinism, from the realm of space, forms of
beauty spring into being under the painter's pencil,
grand symphonies awake under the musician's touch,
immortal verse takes form of life in the poet's fancy,

while those who under their guidance seek out the "old paths" of justice and virtue and love and walk therein may listen to sweet whisperings of rest and peace.

This all-pervading harmony is the standard of truth for the universe, and if other worlds are peopled by sentient and intelligent beings their feelings, tastes, and even moral laws must be substantially the same as our own.

It is fair to infer from the premises assumed that the character of feelings and emotions, as being pleasant or painful, depends upon and is determined by the character and mode of combination of the undulations that compose them. The gentle and harmonious undulations are supposed to constitute the pleasant, while the inharmoniously combined compose the unpleasant feelings and emotions.

Pain in memory ceases to be pain as the undulations which are the cause and record of it in the mind become slower, more gentle and regular with time. If we are aroused to harsh feelings toward our fellow-men, feelings of anger or resentment, our experience nevertheless teaches us that there will come a time when the wave groups in memory will be more in harmony with the conduct we were led to condemn than with our own at the time when the fancied wrongs were fresh.

We therefore anticipate and pursue that course that we know nature will in the lapse of time crown with the approval of our own conscience. We therefore repent or we forgive. All harsh feelings and emotions representing violent or abrupt disturbances of consciousness must in the very nature of things, if left to themselves, take on a more congenial character, and feelings of hate, as the undulations that give rise to them lose their harshness and their character of disorder, are changed into forgiveness.

It is a common lesson of history that the party of mercy is ever the one that in the end gains the world's approval. " Blessed are the merciful; for they shall obtain mercy," is a precept accepted as divine ; and one who read the book of nature with the eye of a seer has told us that

> "The quality of mercy is not strained ;
> It droppeth like the gentle dew from heaven."

Indeed, if hate were not fed with food of unending wrong it would fade from among men. It is evanescent in its very nature. And to pretend that there is a being of infinite power and wisdom who yet can indulge eternal hate, is to pretend that a being of wisdom and justice and love eternally violates laws whose author he must be in order to be at all.

CONSCIENCE.

What else is conscience, then, than this orderly operation of ether vibrations with respect to the affairs of life and to conduct in general?

We have seen the vine, while trying to climb the wall with the help of its tendrils and finding nothing it could grasp, expand the tips of its tendrils into suckers, and with these proceed to perform its duty of offering its flower-buds to the sun.

Can any one doubt that in the protoplasmic mass of its cells, which is the seat of its life, there existed an instability, an erethism, a sense of uneasiness, according to the plant-standard of such things, that was relieved when the vine had obeyed its impulse and the sucker had been substituted for the tendril?

Lower animals of many kinds, after having inflicted injuries on others or on men, have given every evidence of a sentiment of regret.

The behavior of gregarious animals toward each other, as indeed the whole comity of animal life throughout nature, is based upon the operation of a principle that is essentially conscience.

Animals are necessarily prompted by some monitor, indicating the spirit they are to manifest toward their fellows, and that same monitor either approves or disapproves their behavior.

The musician hangs with rapt attention upon the harmony, the melody, the truth of his music. To him discord is pain, is a species of sin, while a vivid pleasure is afforded when the right notes are found and the harmonies attained.

So with that higher harmony, that loftier music whose keys are touched by the fingers of the ether waves, and to which the soul in search of happiness must keep time submissively.

The broad, perpetual, eternal operation of the ether waves make everywhere for kindness, for peace and order, a seeming departure appearing only on the by-paths of selfish interests growing out of the necessities of individual and race preservation.

Sin and wrong and violence and hate are but the

little counter-currents of self in the vast river-flow
of peace and love that sweep forever through the
universe.

Conscience is an ether lesson taught in greater
or less completeness to every living thing.

THE RELIGIOUS FEELING.

We have assumed thus far that all mental affections are due to undulations in the brain aroused
or produced by others from without. We have also
briefly considered the probability that not all the
undulations from external sources find corresponding ones in the cells of the brain. We will now
pursue farther the suggestion that there can be no
response to undulations from external sources except
where the internal structures are attuned to the
vibrations they receive.

If a number of tuning-forks be placed in position
in a room, and another be set in motion, or sounded
in the same room, all the tuning-forks in the collection that are of the same pitch with the one sounded,
and no other, will likewise be set in motion.

Likewise, if a number of clocks be set on a shelf,
and the pendulum of one of them be set to swing-
ing, all the other clocks of the same length of pendulum will likewise start to running, while the others
will not be affected. Such facts lend an air of reason
to the inference that all vibrations that reach the
brain through the senses are lost except those that
find vibrations there of an amplitude and intensity
fairly identical with their own. Again, among the
revelations of the spectroscope is the fact that exactly

the kind of light absorbed by any substance while in a state of vapor will be given out by that substance when made self-luminous.

Sodium, for instance, when in the condition of vapor absorbs only the yellow rays of light, and when made self-luminous gives only yellow light. It is not then a violent assumption that in the process of decomposition organic substances also give out the same forms of force that were absorbed by them from the sun and other luminaries during their growth.

It is these undulations that must be in the brain and perform its work from the time of its first active existence. They are there before the young being has experienced sight or hearing, touch, smell, or taste. And it may be, and probably is, only these undulations, gathered and incorporated by our food during its growth and given out in the nutrition of our tissues, that are re-enforced and shaped into thought and feeling by others coming through the senses from without.

Now there enter into certain of our ideas and emotions something that may not consistently be ascribed to any of the influences hitherto referred to as operating in our minds. We may confidently assume that whatever be the origin of life, we now witness no miracles. Every animal in every part, every leaf in its pattern of shapeliness, every flower with its charm of fragrance and beauty, every fruit in its richness and flavor, whatever may have been the beginning, is now built up and developed by the forces of nature playing on it chiefly from the worlds beyond. It is the little waves of ether coming mostly from the sun that build up the plant, and by their ceaseless pelting drive every atom and every molecule to its place.

Animal life derives nearly if not all its nutritive support from the plant, and thus directly or indirectly from these same ethereal waves, that are light and heat and chemism and the like, all of man except perhaps the mysterious soul that animates him must ultimately be derived.

But we get light and its kindred forces from other sources than the sun, and, pursuing the inquiry, we shall learn that if the worship of the sun was an unconscious recognition of the power it exerts in the maintenance of life, those who placed their deities among the stars may have approached still nearer to the proximate source of the tenderest feelings that contribute to the pleasures of earthly existence, namely, religion and love.

For, aside from any special doctrine of religious experience and of fancied revelation, a religious feeling pervades the human race. Not only so, but a similar feeling is, without doubt, experienced by lower animals. The multifarious theologies that dot history and geography are but the outgrowth of, or inventions made to gratify, the craving of the religious feeling, planted from the beginning in every human breast.

Now, it has been conjectured that every inch of space in the whole vault of the heavens is occupied by the surface of a sun, and that somewhere in infinite space the rays of light must be dissipated and lost, since otherwise the entire vault must needs be a solidly luminous expanse. And, furthermore, that the rays of light and heat and their like are lost to all human sense somewhere in measureless space is a conclusion that would naturally follow from the laws by which their transmission is probably governed. For since, as we have seen reason to believe, though it may not be as yet borne out by direct observation, the undulations of which radiant force consists grow weaker as they advance through space, they must somewhere become too weak to produce impressions as light or heat. While, then, these forms of force, with such others as may have the power of producing sensations or otherwise affecting the mind, are subsiding below the point of intensity at which they may produce sensations, might they not still impress the soul through the apparatus of mind as the gentlest and most agreeable of all the influences it is capable of receiving?

Might it not be that the light from the farthest stars, softened during its immeasurable journey, has become organized in the food we eat, to be given out in nutrition for the building up of cells, and capable there of responding in kind to brother waves coming directly from the original source?

If this be true, then as naturally as the sun and stars give out their light would the mind dispense this store of gentle forces, and most naturally too in the way in which it came, ''For love and love only is the loan for love.''

Or back toward its transformed and personified author the soul would direct it, the very essence of love—Religion! Dominated and prompted by such influences, it is little wonder that men in all lands and in all ages are led to the choice of deities who might be regarded as the adequate source of all that is gentlest and best in their minds and hearts. Influences exerted by the stars may then have much to do in directing the course of life on earth. Their smiles beaming from the skies can not be altogether lost.

"Canst thou bind the sweet influences of the Pleiades?" are words Job puts into the mouth of Jehovah himself. There is not in the brilliance of the sun that which can produce the most pleasurable feelings we are capable of experiencing. Pleasant reveries, tender wooings do not seek his glare, but court the starlight rather. If all this be true, the sun is most probably not the instrument employed in calling the gentlest of all our feelings into being, for as upon an axiom we may rest in the assumption that there can be no evolution without an equivalent antecedent involution.

No creature is greater than its creator, no fountain can rise above its source. The tenderest feelings, then, the heart can know must have a higher origin, a gentler cause, than that of the familiar forms of force; and nothing appears as their proximate source except the fading undulations of light as they journey through infinite space—the "sweet influences of the Pleiades."

THE PHILOSOPHY OF EMPHASIS

—OR—

THE PHYSICAL BASIS OF VOCAL INFLECTION

INTRODUCTORY.

The following incomplete discussion of the nature and cause of emphasis, strictly speaking, forms a part of another essay, ''The Philosophy of Memory.''

But finding that it could not be at all satisfactorily treated in that connection without making a somewhat clumsy episode, it has here been taken up separately at some risk of repetition.

The discussion of the subject is almost necessarily cursory, the aim being merely to indicate the principle, without entering with any fullness into details. Indeed, it is a subject that can not be treated satisfactorily unless it be done orally, since hardly a tithe of the modifications of the voice and their meanings have ever been indicated in writing.

THE PHILOSOPHY OF EMPHASIS.

THE child from its first attainment of the power of speech is able to express its feelings and desires by emphasis and inflection in a way that, in so far as its understanding goes, is not surpassed in the highest state of its subsequent mental development. The rules of emphasis and inflection are of universal application ; for children in all lands, and adults also, where they have not been disqualified by the acquirement of bad habits of expression, observe the same rules and employ the same character of inflection and emphasis for the expression of like classes of feelings and ideas.

No one who has closely observed the expressive modulations of voice among children just learning to talk will contend that the child is under the necessity of learning the rules of emphasis from any teacher, and it is probable that inflection also, as well as every other form of appropriate intonation, is intuitively made and intuitively understood.

Certainly, then, it would seem that we ought to be able to understand that which children come by so easily, so correctly and universally, and to be able to discover and develop the laws by which emphasis, inflection, and other expressive intonations are governed. But in so far as the writer has been able to ascertain, not even an attempt has been hitherto made to develop the philosophy of a class of the most important elements in expression, and one of

the most indispensable aids in the communication
of thought.   It remains to be seen whether or not,
by a diligent study of the problem in the light of
analogy and evolution, we may reach a satisfactory
solution and in some degree subject the phenomena
to a rational interpretation.

First let us inquire why it is that a child empha-
sizes any given word in a sentence.   It is obvious
that he does so, not because he has calculated that
it will impress the hearer more forcibly, or that it
is in any way necessary to speak certain words or
groups of words with more force than others.   To
claim this would be to accord him a degree of intelligence altogether impossible to his years.   He em-
phasizes particular words because they give expression to thoughts and feelings that deeply impress his
consciousness.   He is simply responding to his own
instinctive feelings and promptings while trying to
relieve his mind of an uneasy burden.   But there
must be something behind this still ; and the origin
of the observed effects we must seek in the operation of some yet deeper cause.

We must therefore inquire further what it is that
causes certain words to impress the child's consciousness in such a way that he is led instinctively
to give to them greater force of enunciation than to
others used in the same connection.   In this search
we shall find that the next step beyond this in the
chain of causes is probably the intuitive recognition of the advantage and pleasure of contemplat-
ing objects separately—of having consciousness
occupied with one chief thought at a time.

Even in more advanced life most people realize a pleasure in having one thought occupying consciousness to the partial exclusion of others. Indeed, in the very organization of the mind it is provided that thoughts shall pass through consciousness in single file, so to speak, and that actions shall be similarly conditioned. People eating do not usually like to be interrupted, the less cultivated feeling the interruption more than those of broader mental training, while lower animals feel interruptions most of all. Still the thinker likes to be left alone with his thoughts and the artist with his task.

If we have an object separated and placed out by itself, it is very easy to fix the attention upon it; but while it yet forms a member of a group, this is difficult. If we contemplate a landscape, for instance, we may find in it a vague pleasure, but this pleasure is livelier still when we consider it object by object. At all events the tendency of the mind, the drift of the thoughts, is ever in the direction of the contemplation of objects separately, or of concrete groups that answer to separate objects.

We may see this disposition of the mind manifested on a large scale in history, and in it to a large degree is to be found the source of hero worship. It is much easier to select one who happens to be the most prominent man in an episode and ascribe to him all the merit that is due for some great action than rightly to divide and award his share to each of many participants; for "to him that hath shall be given, and from him that hath not shall be taken away even that he hath," is a maxim grounded in the laws of attention.

One step beyond this takes us probably to the limit of the knowable in this direction. Why this pleasure in the contemplation, disturbed or undisturbed, of objects around us, or the ideas which are their counterfeits?

It has its basis in the love of knowing inherent in the soul, and is of a nature similar to our desire for food and water, our love of pleasant sounds, and our social feelings or love of society. It is the hunger and thirst to know that has become organized within us, and which in all animate things takes the form which we denominate curiosity.

But have we yet reached the end of the contemplated chain that connects curiosity with emphasis? Directly, yes; but there is a collateral link that we must yet consider; and that link is the equivalence and the correlation of mental and physical forces.

Sir William Carpenter long ago showed how mental energy may be transformed into physical exertion, and the doctrine taught by him has become one of the accepted canons of science.

Let us now, in the light of this teaching, proceed to put together our chain of causes and endeavor to get a connected view of the part each element plays in the work of expression. For the purpose of illustration we will suppose a shelf to be occupied by several books to be devoted to different purposes.

We wish to sell one, it matters little which; so we gently and indifferently take one and separate it far enough from the others to be contemplated singly by the prospective purchaser. There is one, however, say the fifth in a line, that belongs to another lot and has been wrongly put with this. If we had to separate this physically from the others we would effect the task by picking it up, removing it with a slight show of force from the others, such a show of force as would be needed to overcome the very slightest resistance. But if some one had put it back in that place after it had been once removed by our order, the show of force would be so decided as to be sufficient to overcome a considerable resistance, in addition to that involved in the mere removal of the book. This would likewise be the identical course pursued by a person deaf and dumb.

If, now, instead of handling the books ourselves, we proceed verbally to order some one else to do so, we would use tones corresponding to and evidently intimately related to our own actions in performing the task we are commanding.

If the contemplated action related to the fifth book, the word fifth would be emphasized in proportion as the desire that it alone should be removed or dealt with occupies our attention, and therefore is desired to be impressed upon the attention of others. If, now, some one to our knowledge has given orders for the removal of the third, we must indicate by our tone in the emphasis of *fifth* not only that the fifth is to be taken to the exclusion of the others, but we must add sufficient force to the emphasis to act as a counterpart or a parallel to the physical effort necessary to resist the attempt made by the intruder to have the third taken instead. The words, in their tone and force, are probably in every case the correlate of the corresponding action.

ANTITHESIS.

A basis of emphasis, as very commonly recognized and extensively employed, is the importance given to each of two or several objects by mentally contrasting or comparing them, and which is known as antithesis. Later on we may find this has its source in the unconscious recognition of the fact that the action and reaction of forces are equal and opposite.

There are two modes of expressing emphasis vocally, one by stress of voice and the other by quantity, and both have their physical correlates; both can be expressed by action.

The less intense forms of emphasis are expressed by stress of voice, the more intense by quantity. Emphasis is not necessarily conveyed in words, nor is the strongest emphasis conveyed in loud tones. The weight almost too heavy to be moved is the one that elicits slow and deliberate effort, and the feeling strong enough almost to defy and baffle expression enlists the whispered monotone in its stubborn utterance. With what slow and measured monotone must Newton have emphasized his regret when he found that his little dog had destroyed the papers that represented two years of his most arduous labors: "Diamond! Diamond! thou knowest not what mischief thou hast done!"

The grating, incisive tones used by men toward each other when about to enter into deadly conflict are seldom loud or marked by deep stress of voice, but often uttered with much quantity and in an almost whispered monotone.

As intimated before, animals that have neither word nor voice are capable by modes of physical movement of expressing the most forceful emphasis. And, curiously enough, these cases present the most apt illustrations of the correlation of the mental and physical forces. Take any of the winged insects armed with a sting, the honey bee, for example. When about to attack an intruder its flight toward the object of attack is steady and deliberate, and the motion of its wings so slow that the insect looks much larger than it really is. Many have thought that the purpose of its slow flight was to allow its wings to be seen in order that it might appear larger and more formidable than if moving with ordinary speed.

But when one reflects that men also approach each other deliberately and somewhat slowly when about to engage in a bitter struggle, and that this is the rule with all or nearly all animals when angered, a better explanation appears to be suggested.

Since mental tension is the correlate of the physical exertion, and capable of being transformed into it, if the animal were to advance to an attack at the most rapid pace possible to it, such, for instance, as it retreats with when put to flight, it would lose the courage required for the attack. It is probable, also, that the lion lashes with his tail, the horse backs his ears, the boar reverses his hair and raises his bristles, and the rattlesnake rattles, not so much with deliberate intention to frighten the adversary, but because such movements are somehow inseparable from the effort to restrain physical action until there is accumulated a sufficient degree of mental and physical tension to prepare the animal for the impending conflict. They are but the accidental overflow of energy accumulating in the form of courage.

The animal is literally nursing its wrath to keep it warm.

This is not a contradiction of Darwin's conclusion that some of these acts have proved of protective advantage to the animals resorting to them, or rather upon whom they have been imposed by reason of the intimidating impression they make upon the adversary. If the horse backs his ears seemingly to keep them from being bitten, if other animals take on an aspect in anger that is of advantage to them in conflict, it does not necessarily militate against the conclusion that the origin of these acts was in the greater necessity of securing through mental tension the proper courage.

A still deeper analysis, if possible to be made, would doubtless show that their ultimate source is in the class of mind vibrations appropriate to each act with which they are strictly correlated.

INFLECTION.

We now come to another mode of conveying variety of meaning, the form of emphasis known as inflection. Here, too, the mental state indicated by the form of vocal expression has an equivalent form of physical expression, and both are probably based on a deep underlying law governing the operation of force.

Thus a demand or an unfinished action would be expressed by one deaf and dumb, or by an animal, by an attitude or movement of aggression or confident expectancy, such as holding up the head or leaning forward. A concession or yielding, on the other hand, would be indicated by an attitude of relaxation or shrinking back. So when a vocal demand is made, or a question is asked that requires a direct answer, the rising inflection is used because this indicates aggression. On the other hand, when an answer is given which is intended as a satisfaction to the question, or a request is granted, it is felt to be an end to the effort, and the falling inflection is employed.

This holds good in some form in every grade of animal life. Any one who has chased cattle, especially the mischievous boy who has chased young cattle till they have become exhausted, knows how, when overcome, they give a peculiar bellow with the falling inflection that announces the end of resistance and their submission to the will of the pursuers. It may be observed in the deadly struggle of foxes, raccoons, and other courageous animals with dogs. When first caught all their cries have the rising inflection, indicating anger, defiance, and aggression. But when all their mental energies, all their powers of resentment and of heart inhibition have been overcome, they cry out with a wail having the falling inflection, which plainly indicates that the struggle is over; the *sauve qui peut*, as it were, of hopeless defeat and utter despair. Even the fierce tiger, among whose progenitors this

sound may never have been uttered since the first
evolution of his royal race, if fought to the death
by an antagonist of overwhelming power, would
doubtless yield the battle and his life in this de-
spairing wail with the falling inflection.   Beasts and
men alike do not need to learn it ; it comes to them
with the forces of nature that organize their being,
and it is probably the same for all animal life in the
universe.

But what of the inflection known as the circumflex?   Here there seems also to be a subtle and
unconscious recognition of the equality of the action
and reaction of forces.

Every capable writer and every thoughtful reader
recognizes the requirement of a certain balance in
the members of a sentence—one part must answer
to or weigh with the other.   If a sentence, either
simple or compound, is wanting in this quality, the
mind is left with a feeling of uneasiness or dissatis-
faction.   It is not necessary that this balancing shall
consist of an equal number of expressed words or
ideas in each of the reciprocal parts.

In one word may be implied the significance of
many, and in one word by this form of inflection
may be expressed the significance of many.   Consequently one word may be sufficient to balance the
weight of many on the opposite side.   The idea
may be illustrated by comparison with the action of
balances or scales when used in weighing.   When
the arms of the balance are equal in length, the
weight put upon one arm is counterpoised by an
equal weight placed upon the other.   But if one

arm is long and the other is short, it becomes necessary to add a greater weight to the shorter arm to
balance a lighter one on the longer arm.

So with the circumflex. Thus, in the sentence,
"I have always befriended you, and yet you will
not now befriend me," the two parts of the sentence balance, and we have merely the emphasis
of the antithesis. But if one should say, "I have
ever been your friend, I have stood by you in need,
I have helped you in want, I have defended you in
danger, yet now when I need your assistance you
desert me," the word "desert" must here have the
circumflex strong enough (the short arm of the
scales must be tipped with force enough) to counterbalance all the parts which go to make up the
preceding part of the sentence, and which answer
to the weight on the long arm in the illustration.

So in the words, "Judas, betrayest thou the Son
of man with a kiss?" we may suppose the force of
the rising circumflex to have been exhausted in the
aim to balance on the short arm of the scale the
history of three years of association, the character
of the victim, and the baseness of the means of
betrayal, all of which were implied in the circumflex and assumed to be present in the minds of both
the actors.

Likewise in Cæsar's dying rebuke, "*Et tu, Brute!*"
the three words counterbalance a lifetime of history
expressed by implication and supposed to be present in the minds of both, pronounced, as they doubtless were, with the fullness of the falling circumflex,
as the lion of so many bloody fields lay down at the
foot of Pompey's pillar.

THE

FUNCTIONS OF THE FLUID WEDGE

— OR —

THE PHILOSOPHY OF SPHERE·FORMING

INTRODUCTORY.

The principle involved in this essay first suggested itself to the author while engaged in studying the origin and uses of the contraction ring of the uterus in connection with the mechanism of labor.

With time and study the field of its application grew, until finally the subject attained its present dimensions.

In connection with this essay the author had prepared another intended as a popular exposition of the question of the tides, in which was offered a supplementary theory attributing tides in part to the reflux of tide masses after the sun and moon had raised them up, and then, leaving them behind, had allowed them to fall back.

The theory, though it seemed to explain some obscure phenomena, was never satisfactorily intelligible, and after the reading of the masterly popular exposition of the tides by Professor George H. Darwin the article was withdrawn, though it had already been placed in the hands of the printer.

The principle of the fluid wedge, in so far as the author is advised, is here for the first time presented, but it is offered in the full confidence that it will in all essential respects commend itself to reason.

THE FUNCTIONS OF THE FLUID WEDGE

OR THE PHILOSOPHY OF SPHERE-FORMING.

IF the finger or a rod be thrust into a glass filled with water, the water will rise up and flow out over the brim. Now the water thus displaced is not lifted up directly, but only indirectly, and if indirectly, it must be by one or more of the mechanical powers. By which one of the mechanical powers, then, is the fluid raised?

Strictly speaking, there are but two mechanical powers, namely, the lever and the inclined plane. The lever, besides the different forms it takes as such, is also modified into the wheel and pulley, while the inclined plane is modified into the screw and various forms of wedges.

In the elevation of the water in the vessel in the case cited there is clearly no leverage involved, and a little reflection will show conclusively that the elevation is effected by means of the double inclined plane or wedge, which wedge in this case is fluid or liquid.

Now, any given mass of fluid or liquid may, as relates to the transmission of pressure, be regarded as consisting of an indefinite number of wedges, extending in all directions and moving upon each other as solid wedges without friction.

Let us take, for example, an inclosed cubical mass of water (Fig. 1) and conceive it to be divided into two equal wedges, A and B.

It is quite obvious that as regards the transmission of stress or pressure the two wedges will behave exactly as if they were both solid and moving upon each other practically without friction. For whatever support A may receive from B and

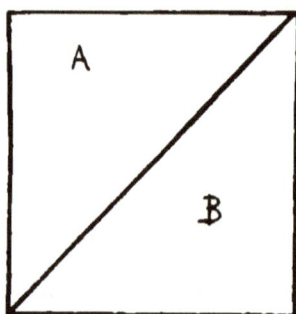

Fig. 1.

the walls of the containing vessel will be counterbalanced by an equal support that A will return to B and the walls of the containing vessel, and reciprocally; each wedge will therefore have the functions and force of a solid wedge.

WHY FLEXIBLE TUBES BECOME ROUND.

If now, instead of a cubical vessel, this water is inclosed in a flattened, flexible tube, such as a piece of rubber hose, for instance, and we conceive the liquid in the end of a section of that tube (Fig. 2) to be divided into wedges with their bases of uniform areas and resting on the wall of the tube, while their apexes meet at the center, it is evident that

the wedges A and B laid off in the largest diameter
of the tube will be longer, more slender, and therefore more acute than the wedges C and D laid off
in the shortest diameter. It is also obvious that
whatever support any one of these wedges might
derive from the other wedges into which the section
of the tube is divided will be yielded back by it in
return.

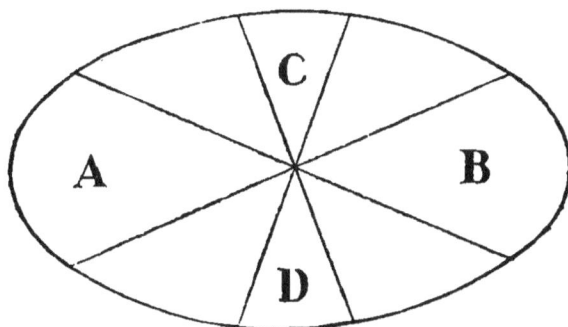

Fig. 2.

Let us now suppose pressure to be applied
equally to the bases of all these wedges, which may
be done by forcing water into the tube. Since A
and B are more acute than C and D, they will
advance more easily under a given stress or pressure, and consequently C and D will be forced back
from the center, their bases carrying before them
the flattened sides of the tube wall, until all the
wedges become of equal length and the tube
becomes cylindrical.

WHY THE SOAP BUBBLE IS A SPHERE.

The tendency of air bubbles to take on a spher-
ical form must have been among the earliest observed physical facts. Almost equally familiar must

have been the tendency of liquids to form spheres when their particles are left free to exert their attraction upon each other. No explanation of these facts, however, based upon definite principles and a final analysis 'has ever yet been given, in so far as the writer is aware, unless the mathematical demonstrations based on surface tension might be called such. On the hypothesis of the fluid wedge the explanation becomes simple and easy.

But now, instead of the end of a flattened tube, let the previous drawing (Fig. 2) represent a section of a closed sack of such construction as to be susceptible of being extended into a sphere, but only partly or laxly filled, and therefore somewhat flattened. The inclosed wedges will now be replaced by cones, with their bases to the wall and their apexes to the center of the mass of fluid.

As with the wedges, these cones will also mutually support each other and move upon each other practically without friction.

If we proceed to subject these cones to pressure by forcing fluid into the containing envelope—say a rubber toy balloon—the longer cones or wedges A and B will advance more readily than C and D, and since all the longer or more acute cones or wedges will in a proportionate degree advance more readily than the shorter or more obtuse ones, the latter will be driven back and will carry the wall of the containing envelope before them until all become of equal length and equal volume. The fluid with its envelope will then have become a perfect sphere, and all its parts will be in equilibrium.

WHY RAINDROPS FORM SPHERES.

If, however, instead of a pressure from without upon a fluid confined within a containing membrane, we assume the fluid to consist of a mass, free in space, but subject to the operation of the mutual attraction of its particles, all these particles being drawn thereby in the direction of this common center, then the same result of sphericity will be reached as before.   Physicists will readily carry this analysis a step further and perceive that such behavior of the wedges is due ultimately to the direction of the lines of force, these lines being directed always at right angles to the surface of the inclined planes or wedges and cones.

Furthermore, it is necessary to conceive that these fluid wedges do not actually advance or recede in mass, as solid wedges must do, but that they do so in effect by a kind of decomposition and recomposition, the molecules shifting from the acute and more massive wedges and becoming part of the smaller obtuse ones until all are equal in length and volume.

BEARING OF THE PRINCIPLES ON THE FORM OF
THE EARTH.

Without resorting to the formal analysis of the principles of sphere-forming here attempted, it has been very generally realized that a liquid mass, left free to the mutual attraction of its own particles, must form a perfect sphere.* In examining the contour of the earth, however, we find many departures from this form, aside from the shortening of its axis due to revolution. The principal of these departures are the enormous mountain chains and the abysmal depths of the sea. If the core of the earth now consists of a plastic mass, be it as rigid as a plastic mass can be, the earth must be in equilibrium, and a cone with the top of the highest mountain for its base and its point at the center of the earth must weigh as much as another with the same diameter of base taken from the bottom of the deepest sea. If it be said that the mountains are supported arch-like on the crust of the earth, it may be answered that they had no such support when they were lifted up. They were plastic then, and if out of equilibrium should have settled down before they cooled and had need of the support of an arch. It seems almost certain, then, that the specific gravity of the sea-bottom crust is greater than that of the dry-land crust.

THE BALANCING OF FLUID IN CONNECTING CHAMBERS.

Many more people have wondered than understood why the water in the spout stands on a level with that in the body of a teapot. In this principle of the fluid wedge we have a ready explanation of the

---

*The suggestion is made by Major Wm. J. Davis, of Louisville, that, given an abundance of atmosphere and water, a rotating body, no matter how rigid nor how irregular in shape, would eventually take a spheroidal form ; the higher parts being worn away and the lower ones raised by the movement of eroded material.

definite process by which a liquid in even the smallest tube assumes the same level — barring capillarity — as that in a large vessel with which it may be connected. It affords also the basis of a clear insight into the principle of the hydrostatic press.

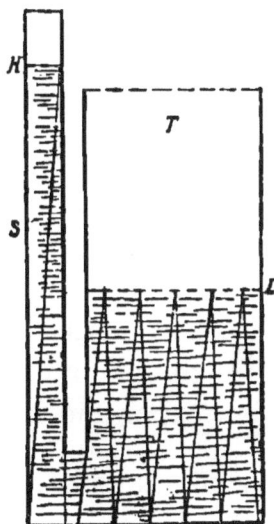

Fig. 3.

In the drawing (Fig. 3) let T be a tank filled with water up to the level L and connected with a tubular arm S in which the water stands at the height H. Now let us conceive the water in both the tank T and the arm S to be divided into wedges having bases of equal area and extending from the free surfaces to the common bottom. Obviously the wedges into which the water in the arm is divided will have greater length and will be more acute than those in the tank. It must follow, then, that owing to this greater acuity and greater angle of the lines of force, any given strength of blow or given pres-

sure on one of the acute or long wedges in the arm
will cause it to advance more than one of the obtuse
wedges in the tank will under an equal blow or
pressure ; that is, the slender wedge will be more
easily driven than the blunt one.   And, for all practical purposes, blows are being struck continuously
or with a rapidity practically infinite upon the bases
of all the wedges.

Now, since in our example the rapidity of the
blows or the amount of pressure is the same upon
the base of every wedge, it follows that the acute
wedges in the arm will be advanced with less resistance than the obtuse wedges in the tank, and the
obtuse wedges will consequently be forced to recede
until the liquid rises to the same level in the tank
and its arm — in the teapot and its spout.

NATURE OF LINES OF FORCE.

It may not be amiss at this point to digress for a
moment and try to make plainer to the non-scien-
tific reader this matter of the lines of force.   We
have seen that lines of force are extended at right
angles to the surface of an inclined plane, and therefore at right angles to the two surfaces of a wedge,
which is a double inclined plane.

An idea of the action of lines of force may be
gained from the assumed construction of stools out of
a number of top-shaped pieces of wood or wooden
cones and an appropriate number of legs.   That is,
the legs must be proportionate in number to the
taper of the blocks or wedges, so that the greater
the slope the greater must be the number of legs.

Thus the stool, Fig. 4, would have a great number of legs, and it would offer great resistance; that is, it would support great weight. The stool, Fig. 5, on the other hand, would have few legs and

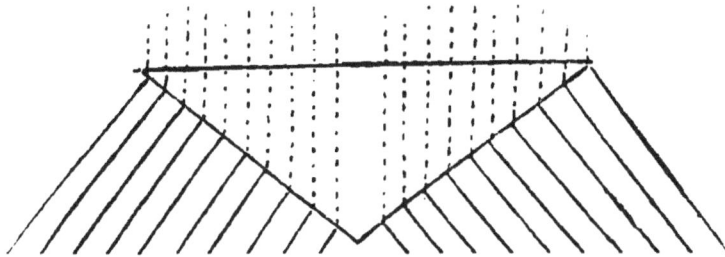

Fig. 4.

they would spread out widely. The resistance offered by these legs to a downward pushing force would not be great, the stool would not be strong, and it would not therefore sustain a heavy weight. If the body

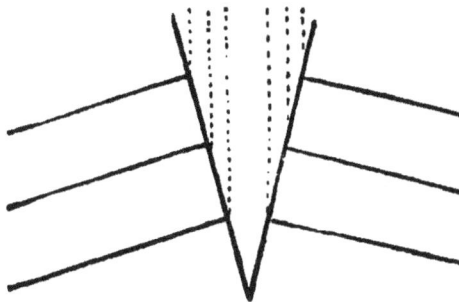

Fig. 5.

of the stool had almost no taper, it would have almost no legs, and what it might have would point almost directly outward instead of downward; and every one's experience tells him how easily such a stool could be crushed.

Now, let us conceive the force opposing the
advance of a wedge into the trunk of a tree, or even
a mass of water, to be divided into units, each one
of which points, as did the legs of our stools, according to the taper of the wedge. It is clear that the
wedges having much taper and having the lines of
force most nearly parallel to their axes will meet with
more resistance than those having fewer units of
force, with the lines of force pointing at a larger angle
to the wedge axes.

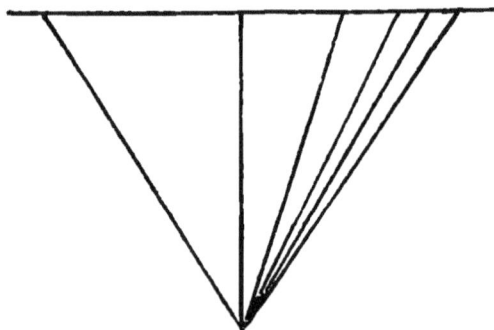

Fig. 6.

And now, instead of a stool, let us take a blunt
wedge, and, instead of the legs, let us conceive the
resisting force divided into units or rays ; and let us
also conceive this wedge to be split successively into
smaller wedges, as in the drawing (Fig. 6).

It is clear that our rays of force for each wedge
so formed decrease in exact proportion to the
increased slenderness of the wedges. If these
more slender wedges have only a millionth part of
the slope of the obtuse one from which they were
formed, each one will have only a millionth part of
the rays of force resisting its advance by bracing it

at right angles to the incline of its planes, which in this case will be nearly at right angles to the axis of the wedge.

This process may continue until each wedge becomes infinitely acute, when it will have resisting its advance only an infinitely small part of the force resisting the advance of the original wedge.

If the wedge has its planes nearly parallel and is almost without slope, it will enter between and separate two bodies with little resistance other than that offered by friction. A cone or wedge of the width of base of half a mile would, under equal pressure of equal area, exert no stronger lateral pressure than one of the thickness of a needle, both being of equal length; for each would return to the rest of the mass of water just as much as it received from it. The total lateral resistance of all fluid or liquid wedges, therefore, is the same, whatever may be the diameter of their bases or whatever the incline of their surfaces, provided they are of equal height.

This brings us, then, to an explanation of how a column of water in a tube no larger than a straw can raise the body of the ocean — could raise a mass of water almost of infinite proportions. We first conceive the ocean to be divided perpendicularly into wedges, one half of them with their bases upward and the other half with their bases downward. These wedges, though half a dozen miles in length, might have their bases and apexes to correspond in thickness to the two extremities of a cambric needle.

Furthermore, the most slender column or wedge conceivable extending from the free surface to the bottom of a body of water will exert as much lateral force from the position it occupies as all the rest of the body, even if it be the ocean, can exert on it. For to each and every other wedge or column it returns as much as it receives. Here, again, the principle is the same as with solids. If a tree were split through its whole length into fibers, these would scarcely more stand upright unsupported than would like slender columns of water. In this, as in other respects, liquids are but other forms of solids, as solids are but other forms of liquids.

The lines of force with wedges of incline so small as those described would be directed practically at right angles to the axis, and therefore practically no resistance would be offered to their direct advance. When, then, any such wedge should be found of greater length than the others, having from this cause greater weight with no more resistance to its advance than the others, it must continue to advance or sink down by the process of decomposition and recomposition until it should become of the same length.

While a wedge of water is thus settling down into the sea, the water of the sea may be assumed to be divided into horizontal wedges of indefinite length, and these would be pushed apart to raise the level of the rest of the ocean.

NO HYDROSTATIC PARADOX.

In the action of the fluid wedge we may also find an explanation of the so-called hydrostatic paradox, or the law, as commonly stated, that water presses equally in all directions, and that a quantity of water, however small, may be made to counterbalance any weight, however great.

Let us suppose that a small tube with its contained water divided into wedges rises above and is connected with a closed cask of water. Now, it will make no difference in the application of the principle whether we conceive the water in the tube to form wedges extending down to the bottom of the cask, thus forcing the horizontal wedges into which the water in the cask is resolved laterally upon each other, or whether we regard them as bending with thin extremities on entering the cask and extending out under its head. In any event, all the wedges at the same level will be equally affected by pressure, and the driving out of these horizontal wedges lifts the head of the cask, even it may be to bursting it off. The fluid wedges in this upward pipe have done no more, have exerted no more power than solid wedges would have exerted in a cask of other solid wedges could they all move upon each other without friction. There is then no hydrostatic paradox — the paradox disappears.

DISPLACEMENT OF FLUIDS.

The explanation of the reactions involved in the displacement of fluids would seem to be easier on this principle than by the method usually employed.

Thus, if a cylinder C of cork or any light substance one third of the diameter of the vessel V, partly filled with water, be let down into it by the cord R, it will force down the wedge A, which will slide upon B and lengthen the wedges 1, 2, 3, and 4 on either side of it. Now, the farther down into fluid the cylinder C is forced, the more obtuse will

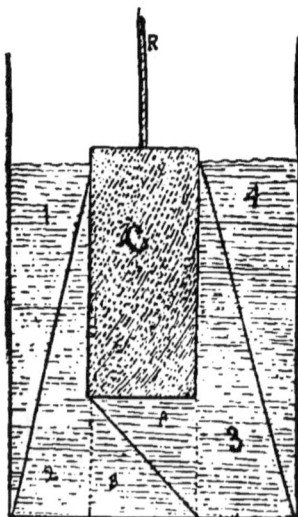

Fig. 7.

the wedges A and B become, the more resistance will be offered to their further shortening, and consequently the greater will be the reaction by which they resist or force upward the cylinder. This, of course, will be due to the increasing preponderance of the wedges 1, 2, 3, and 4 over A and B by reason of their increasing relative acuity.

If instead of the cylinder C a mass of material such as may be submerged in the vessel and yet lighter than water be employed, the resistance to its deeper immersion will not increase after it passes beneath the surface, for the water moving in above it will equal in weight that displaced.

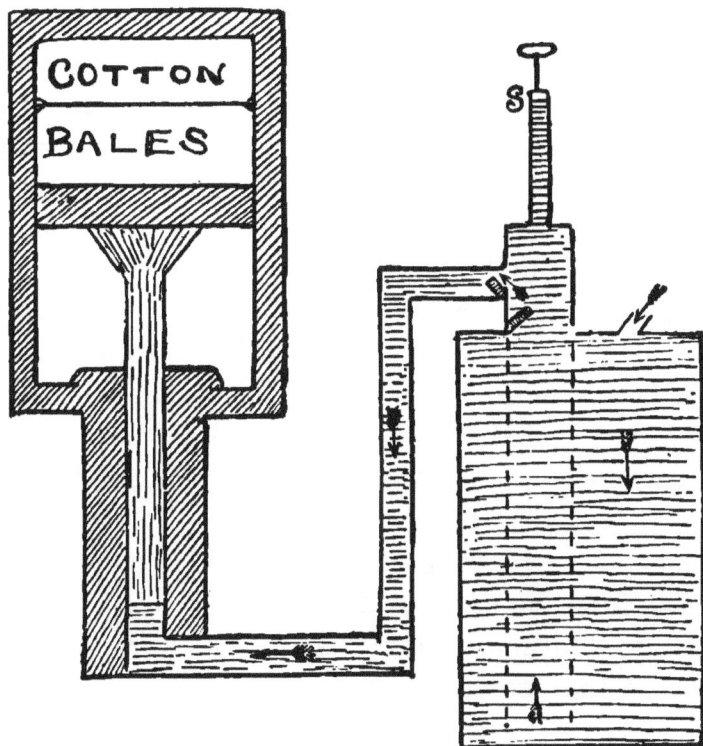

Fig. 8.

HYDROSTATIC PRESS.

In the hydrostatic press identically the same principle is involved as in the balancing of the liquids in connecting chambers already described. The water in the tube extending from the plunger is supposed to be divided into wedges of the same length as those in the tube. A single wedge, how-

ever, the diameter of whose base would equal that
of the bottom of the plunger or piston, would conduct all the force, and would be the unit of measurement. When the press is in operation every
potential or hypothetical wedge in the piston chamber will have a force exerted on it equal to that
exerted by the plunger in addition to its own weight,
and all of them will have a total force exerted on
them equal to the force exerted on the wedge in
the tube of the plunger, multiplied by the number
of the wedges of equal area of base contained in
the piston cylinder.

The water in the piston chamber and tube has
been so far treated of as if divided into wedges of
the whole length of the space intervening between
the piston and the base of the press. Practically,
however, this is not the real condition ; otherwise
the tubes might be extended indefinitely with the
wedges of corresponding length, and thus an incal-
culable force be exercised.

Under ultimate analysis the wedges are the
smallest that could be imagined to be constructed
out of the molecules of water, and are in fact
ultramicroscopic.

The only length required would therefore be
such as to admit of the formation and action of
these diminutive wedges. All additional length of
the tube, even if it were perpendicular, could be of
advantage only by the additional weight of water it
would supply ; while if horizontal, it would serve
only to increase friction and impair the power of
the machine.

THE BOURDON STEAM GAUGE.

This instrument consists of a bent tube or hollow metal receiver which straightens when steam is admitted to it under pressure, and which has an apparatus attached to it for registering the degree of straightening. The explanation of this straightening usually given in the text-books is, that it is due to the pushing apart of the walls of the tube, which is curved on the flat. This is clearly an insufficient explanation, since it is not at all necessary for a

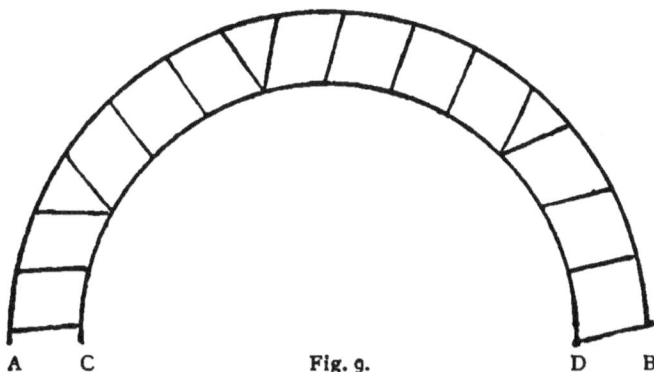

A   C   Fig. 9.   D   B

curved tube to be flattened in order to have it straightened under the pressure of contained fluid. Professor Peddie in his text-book on physics makes the generalized mathematical statement that the straightening is due to the fact that the sum of the moments of force between C and D is greater than between A and B; but this, while mathematically true, gives no definite analysis of the physical principles the case involves. The principle of the fluid or liquid wedge will apply here also, for it applies as well to gases as to liquids.

If we conceive the steam in the receiver of a
Bourdon gauge, or the fluid or liquid in any other
tube, to be laid off in transverse sections of equal
length, we will find that in order to keep pace with
the curve several wedge-shaped sections must be
provided for, these wedges all having their bases on
the convex or long side of the tube, and the apex
on the concave or the short side.   Now, since the
lines of force act at right angles to the inclined
planes forming the wedges, it will result that the

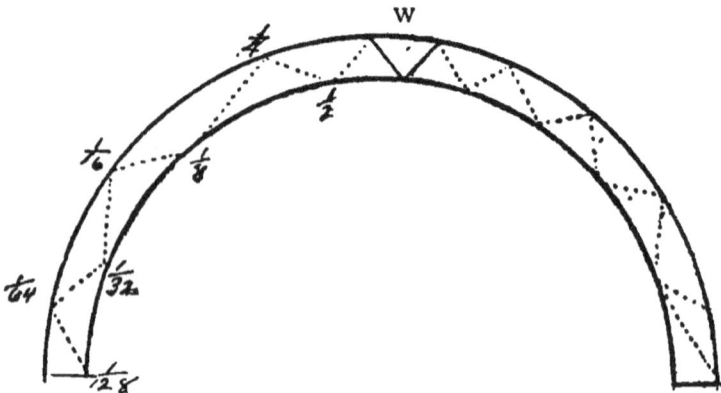

Fig. 10.

rays exert themselves most on the inner or shorter
wall, and cause it to stretch more than the outer
wall, thus straightening the receiver.

The interaction of these lines of force may be
very complex as between the surfaces of the wedges
and the walls of the receiver, but the total result
will be the same as when the total resistance of all
the wedge surfaces is divided between the two walls
of the receiver in such a way as to give the inner
wall the first half, then the outer wall half of the
remainder, and so on through an infinite series.

Thus, suppose the wedge W (Fig. 10), with its base resting on the outer wall of the curved tube, to be filled with fluid, and its point extending to the inner wall to be driven forward by the pressure of the steam forced into the tube. The lines of force will impinge first on the inner wall. According to the law of the deflection of force, one half of it will be exerted on the inner wall and the other half will be deflected to the outer wall. Reaching the outer wall, one half of this half, or one fourth of the total, will be exerted on it and the other fourth will be deflected to the inner wall, where one half of it, or one eighth of the total, will be exerted, and so on until all has been exhausted. Thus, the inner wall having received first one half and then one eighth of the total, has already had exerted upon it the greater part of the moment of force proceeding from the wedge.

Distributing the force exerted by the rule applicable to an infinite series, we have a total of two thirds for the inner wall and one third for the outer wall. It is obvious in this case, as in others, that these injected wedges in the tube will act on the whole exactly as if they were solid, for whatever support they receive from the rest of the contents of the tube and the tube wall they return to them.

Innumerable phenomena involving the displacement of fluids may have light shed on them by viewing them in connection with this wedge action of fluids and gases. The principle has an important bearing on the movement of ships, and it is quite extensively involved in the movement of fishes beneath the water. Thus, a whale swimming deep in the sea might, at first blush, be supposed to be overcoming an enormous resistance. But if we may assume that all the water in contact with the body composes the bases of wedges, extending with their points so far out into the sea that there is comparatively little resistance to their advance, and that the wedges it drives before it are very nearly counterbalanced by the reaction on those behind it, the difficulty of the task largely disappears.

DISSIPATION AT THE MOUTHS OF RIVERS.

At a hasty glance the conclusion would be a natural one that rivers should empty into the sea by an abrupt front or a kind of waterfall. How can the outflowing river push away from its mouth the immense mass of sea water and keep on its fixed march with the ocean column without heaping up or without a single ripple at their place of meeting ? It must be that by wedge action the water of the ocean is lifted far out from the shore, and that by its means masses of water, great as even the Amazon delivers, are in effect stored almost instantaneously in far-off ocean depths. It is most likely, too, indeed it is almost impossible for it to be otherwise, that the surface of the sea slopes up to the mouth of every river in order to give the potential incline necessary to enable the river to wedge off its waters into the ocean.

REFLUX OF TIDE MASSES.

It matters not what theory of tides may be adopted, one conclusion in which all agree is that the elevated masses of water do not follow in identical form the movements of the bodies raising them. Their constituent parts change from moment to moment. The tide mass of this hour sinks in the next into the ocean, and another is raised, made up of water largely different and distinct. Yet, how do these masses disappear, except by far-reaching wedges lifting other parts of the sea, and even, speaking in a potential sense, carrying the water back beneath the moon and beyond it, which the moon had raised up just the hour before?

If the earth were a mass of water, a wave or tide elevation would sink down into it simply as a wedge, and would not travel over its surface. As the case with the ocean actually is, however, the water that is drawn up or thrown up into tidal elevation must be dissipated by wedge action, interfered with by friction at the sea bottom ; and it would seem, also, that the larger the area of the mass elevated, the height being the same, the more rapid would be the movement of dissipation. It would simply sink down into the sea, and the water beneath it would be thrust out in every direction by the wedge action in a wholly different way from that in which a wave is propagated. This, however, remains to be confirmed.

If an enormous mass of water were let sink into an ocean, say thirteen miles deep, would its dissipation await a wave movement of one thousand miles an hour, or would it be effected by wedge action far in advance of any possible wave?

It is a fact, however, that tide masses do undergo dissipation under conditions that necessitate the movement of the water at the rate of one thousand miles an hour, a rate at which it is impossible for waves to travel with the present depth of the ocean. Therefore, if this be not accounted for by the action of double inclined planes or wedges, it is a problem that yet remains to be solved.

THE BIRTH OF A PLANET

A CRITICISM OF THE NEBULAR
HYPOTHESIS

INTRODUCTORY.

Even though the principle advocated in this essay should be shown to be quite inapplicable to the case under consideration, it is nevertheless not without use in the interpretation of other phenomena. Its utility is especially exemplified in searching for the immediate source of tornadoes, which it proves to be an offshoot of the cyclone and a part of the cyclone mass that has been thrown off at a tangent. If this were not the case, tornadoes would necessarily rotate in a direction opposite that of the cyclones from which they spring, instead of the same direction, as they actually do.

There is little pleasure, it must be confessed, and less of profit in disturbing conclusions that have come to be regarded as settled canons of science, unless one is able to offer for the rejected principles something that is at least as satisfactory.

The nebular hypothesis has been so long regarded as meeting the conditions of planetary evolution, and is withal so beautiful and ingenious, that the writer can not avoid feeling that he inflicts loss upon himself as well as others in so far as he may happen to succeed in weakening it. He is therefore quite willing to be proven to be in error in entertaining an adverse opinion.

THE BIRTH OF A PLANET.

A Criticism of the Nebular Hypothesis.

THE nebular hypothesis suggested by Sweden-borg, developed by Kant, and supported by La Place, proposing to account for the origin of the various heavenly bodies, assumes that in the beginning the atoms of which worlds now consist were disseminated through space as a nebulous vapor. For such a system as our sun, it was supposed that this vapor began gathering into a central mass, and, in doing so, inaugurated a revolving motion around a central axis. In the course of time this mass shaped itself into a revolving sphere, or disc rather, with an enormous equatorial diameter and a relatively small axial or polar diameter.

As the mass cooled off, the central parts contracted and gradually withdrew from the outer part or rim, which remained as a ring surrounding it, kept from approaching nearer the center by centrifugal force due to the revolution of the whole mass of vapor.

This ring from some cause moved more slowly in some places than in others, until finally its particles all got together in one mass, forming a planet. In this way one planet after another was formed, until at last the sun reached its present dimensions.

On the same hypothesis it is assumed that when these planetary masses got to revolving on their axes, nebular rings were in like manner left surrounding them, and these in time became satellites or moons.

FACTS IN ITS FAVOR.

The facts in favor of this hypothesis are both numerous and significant. In the first place, from two hundred to three hundred planetary bodies are known, all rotating in the same plane and in the same direction, with the exception of the satellites of Uranus, which by some means have been caused to rotate at a very considerable angle to the common plane of the solar system.

Between Mars and Jupiter are a large number of satellites revolving around the sun, and supposed to be the remnants of a nebulous ring which failed to become a planet and broke up into these smaller bodies.

The probability that such an arrangement as that just described and observed to exist among the planets and satellites of the solar system could have been brought about by a fortuitous aggregation of nebulous vapor, when tested by the doctrine of chances, is almost infinitely small.

Again, it has been ascertained that the sun is now diminishing in diameter by about four miles in every century. If this has been steady and continuous in the past, forty-seven million years ago the sun must have occupied the space the earth does now, and some time in the past must have extended beyond the orbit of Neptune.

Observations by means of the telescope have also lent additional probability to the theory. Astronomers have discovered nebulæ in various stages of progress toward the formation of suns. Some have been seen without nucleus, others have presented a nucleus more or less translucent, while still others have revealed extensive nucleus formation opaque to the light of other luminous bodies.

Another feature that favors the theory is the central heat of planets. It is entirely consistent with the hypothesis that planets should begin cooling off at the surface, as is evidently the case with Jupiter, the earth, and the moon, while the interior mass remains incandescent, as is the case with at least the earth and Jupiter.

DIFFICULTIES OF THE HYPOTHESIS.

Not to mention the irregular orbits of the satellites of Uranus, which may well have had an accidental cause, there are still some other difficulties in the way of the acceptance of this theory; but there is so much in favor of the hypothesis, and mankind has been so completely baffled in devising any other mode of origin for planetary systems except that of direct creation, that the world generally has accepted it on authority.

Yet there is one objection that has not heretofore been offered, or rather it might be said there is one fact that has been offered in behalf of the hypothesis that presents so great an obstacle that it appears to require a material modification in its statement. This objection, or at least difficulty, lies in the fact

that the planets and satellites rotate on their own
axes in the same direction as what is supposed
to be the parent body—the very fact that probably
has been relied on most of all to prove the theory.

MUST BE THROWN FROM PARENT BODY.

This rotation about their axes on the part of
planets and satellites could be brought about only
by their being thrown off to some distance from
their position as a part of the parent body, and not
by the withdrawal of the parent body from them,
as assumed in the hypothesis. The earth, for instance, could revolve on its own separate axis as
distinct from that of the sun, and in the same direction as the sun, only by being caused to move in a
larger orbit than that described by it while it was
still a part of the sun's mass ; but it could not do so
while retaining the same orbit or moving in one of
smaller diameter.

If a cane be held in the hand by one end, and
then let go in a straight direction after being swung
rapidly around from behind forward, it will begin to
rotate as soon as it leaves the hand.

The reason is that when let fly the distant end
was moving more rapidly than the near end, and so,
trying to outtravel it, had to pass around it and thus
inaugurate rotation about the axis. Again, when a
wheel or disc is revolving, the outer parts are moving faster than the inner parts ; and this is true of
every two particles in the disc, where one of them is
nearer to the center than the other.

This character of motion is called linear motion, and the linear motion of every particle is greater than that of any other particle nearer the axis of rotation. But every particle of the disc has also another motion known as the angular motion, and this is the same for every particle in the disc or wheel, no matter where situated.

Thus a particle near the axis may move but one foot a minute, while another at the circumference may move a hundred feet in the same time, but they both pass through the same number of degrees, or the same angle. It is obvious that every point in the wheel must have a fixed ratio of angular to linear motion ; the farther from the axis the greater the linear motion must be for each degree of angular motion.

Now, if a part of the disc at any point should be freed from the mass to which it belongs, and all its particles should preserve the same linear and the same angular motion that they had before separation, it would simply continue to go around in its path, without rotating on its axis in either the one or the other direction. But if it should be drawn closer than before to the axis of the body from which it had separated, it would revolve in a direction opposite to that of the original disc ; while on the other hand, if it should be thrown farther from the center, it would rotate forward in the same direction as the original disc.

Now, let us suppose that a block or fragment from this disc flies off at a tangent while it is revolving. Clearly, since the outside of the disc is moving faster than the inside, the outside of the fragment will be moving faster than the inside, and will of necessity move around the inside and cause rotation in the same direction as the disc from which it was broken. This could, of course, occur only when the fragment should fly off in a straight line, and the equilibrium between the linear and angular motions be disturbed.

Or let us give one more and a homely illustration :

If two horses are running on a circular race-track, the one on the outside must run a little faster than the one on the inside in order to keep abreast of it. The outside one can run a little faster than the inner one without ever running around it. But let them now fly the track at a tangent, running side by side in a straight line, each keeping up the same speed as before. They will thereupon be seen to revolve around each other, the outer running around the inner one in the same direction they maintained while on the track.

Let us now carry out the parallel and learn what must occur with a nebulous ring left in space by the contraction of a central mass. In the first place, it is exceedingly difficult to conceive how all the particles of such a mass could get together.

The orbit of Neptune, for instance, is nearly eighteen billion miles in circumference, and the particles of matter gathering from the nebulous ring to form the planet must, before this task could be accomplished, have traveled exactly in the same orbit, with a uniform, continuous, and independent slowing for all the particles in one direction from a given point, and a like acceleration for all the particles in the other direction.

It would not suffice to assume that any inconsiderable part of the ring became accelerated and hastened around, driving the remainder before it to be massed into the new planet, for the shock of the concussion of the fast-moving with the slower-moving parts would soon arrest the excessive motion.

While, on the other hand, if one part should become retarded or arrested in its motion, unless such part should constitute a very large proportion of the total mass, it would soon again be set into rapid motion, as it would be overtaken by the after-coming mass of particles. If this view, then, is to be accepted at all, it would be more reasonably assumed that each particle experienced its due proportion of retardation or acceleration. This of itself is hard to conceive and harder still to credit.

It takes Neptune one hundred and sixty-four of our years to revolve once around the sun. Light itself would require more than three years' time to make the journey.

How long a period, then, would be required for one part of a nebulous ring to leave the part next behind it, to move around in the orbit, bring up the rest, and overtake that part again? Furthermore, it is to be considered that this is to be accomplished in the absence of any known or suspected cause of acceleration for the one part or retardation for the other.

But it is nowhere suggested in the nebular hypothesis that the rings fly off at a tangent, or that they
in any way recede from the parent bodies; and if
they do not, we have seen that it is not possible for
them, even should they get together in planet forms,
to take on rotation about their own axes. Yet this
must have happened in the case of all of the nearly
three hundred globes constituting our planetary system if we are to accept the current statement of the
nebular hypothesis.

GEORGE H. DARWIN'S MODIFICATION.

Professor George H. Darwin, son of Charles Darwin, and famed as an eminent mathematician, has
made the startling suggestion that the moon, at least,
was lifted up on the earth as a tidal elevation, and
then thrown off into space. Now, this would meet
the rotational difficulty as far as the satellites or
the planets near the sun are concerned, but what
tidal elevations could be produced on such distant
planets as Neptune or Uranus, or even Saturn or
Jupiter?

Even by the time the lines of the sun's attraction
reach the earth they have become so nearly parallel
that the sun can raise only two fifths as much tide
as the moon, although twenty-five and a half million
times heavier. But before they could reach Uranus,
which is more than thirty times as far away as the
earth, the lines of the sun's attraction must be so
nearly parallel that tide generating force would be
practically obliterated, and the tides would become
a vanishing quantity. Now it can hardly be the

case that there are two causes operating in the universe at the same time, both capable of forming nebulous masses into revolving satellites and larger planetary bodies.

But Professor Darwin does not assert a tidal origin for the primary planets. On the contrary, he distinctly declares his belief that the planets have, from the beginning, revolved in substantially the same orbits as at present.

SUGGESTION OF COMETS.

Would it, then, be too wild a dream to imagine the universe peopled with cometary bodies during the first ages, and to suppose that they possibly passed close enough to the shrinking suns to gather up a portion of their mass, and that then, carrying it some distance away, they joined it in revolving around the parent body? A comet striking the nebulous border of a sun in a direction opposite to that of its motion would be checked by its momentum, and might fail to carry away any part of it, but on the contrary would itself be lost in the mass. So if it should strike across the direction of motion, it might fail, boring through and carrying away only a small portion of its mass. But if it should pass through a nebulous border in the direction of its motion at a time when that border was barely restrained from flying off by reason of the fact that the centripetal and centrifugal forces were nearly balanced, it could possibly carry it far enough away to set up a movement of rotation and then revolve with it as a planet around the parent body.

After all, however, one might be tempted to suggest and might be excused for suggesting that worlds have a season to bring forth, as do animals and plants, and that in their proper times and seasons, fixed in the infinite councils, they drop their ripened fruit of young worlds into space.

THE LAWS OF RIVERFLOW

— OR —

THE TRUE THEORY OF STREAMS

INTRODUCTORY.

Of the investigations of the ancients on the motion of fluids, only those of Archimedes have come down to us, though it is not probable that one could have reached the station that he adorned without having had others to share the interest he took and thus to encourage him in his work.

After his time many centuries passed during which not a word of all the records that have reached us tells that men even thought of the streams as a matter of investigation.

Hydromechanics must then be regarded as a modern science which virtually owes its existence to the great men who adorned the seventeenth and eighteenth centuries. Italy may be said to have been its birthplace.

Every point connected with the theory of torrents and rivers, the conducting and distribution of water, the slopes, the directions, and the variations of channels were sedulously inquired into by Castelli, Viviani, Zendrini, Manfredi, Guglielmini, and Frisi.

The study was really inaugurated by Gallileo, the father of modern astronomy. He was followed by Castelli and Torricelli, two of his distinguished disciples who, stimulated by the great interest the question of rivers has for a country so dependent on irrigation and at the same time so liable to disasters by floods as Italy, attempted to apply to rivers the principles enunciated by their great master.

Guglielmini, in a publication made in 1686, came very near reaching the position maintained in the present treatise, contending that the retardation of streams and the regulation of their movement were due to transverse currents at the bottom caused by friction against the rough beds. But Mariotte, an eminent French authority, having ascertained by experiment that streams were likewise retarded in channels made of smooth glass, Guglielmini abandoned his position, and in a subsequent edition of his works tried to account for the phenomena on other grounds. Father Grandi also recorded observations in which he had seen stones carried out transversely against the banks of streams.

When the transcendent Newton flashed upon the world like a new star in the firmament of science and philosophy, he deemed the subject not beneath his attention. He spent much time in its investigation, and even devoted a part of the Principia to problems relating to the movement of fluids in tubes and channels. Since his time a multitude of able investigators have taken their turn at the elusive puzzle. Experiments innumerable have been made, and at least three sets of these, namely, those of Captain Gordon on the Irawaddi, Major Allan Cunningham on the Ganges Canal, and Darcy and Bazin at Paris, make reports of more than two thousand pages each. Major Cunningham made forty thousand gaugings and experiments. Prof. James Thomson, brother of Lord Kelvin, and a man of great learning and ingenuity, has given it extensive study, while Forbes and Tyndall have investigated the movements of glaciers.

In France the study has been diligently pursued for nearly two centuries by such eminent men as Mariotte, already mentioned, Pascal, D'Alembert, Dubuat, Bossut, Bernouilli, Boileau, Darcy, and Bazin. In our own country, Captain Eads and Humphrey and Abbott are among those who have given the subject close and extensive study.

It was not, however, by following up the experiments and calculations of others that the writer reached what he is entirely confident is the true solution.

He was first set to thinking on the subject by a schoolboy adventure. In the early spring of 1858, on an uncomfortably cold day for the season, the author and a fellow-student went swimming in the Ohio near Brandenburg, Kentucky, as the result of a banter. His companion took a plunge and withdrew. But he himself, having rolled a treelap into the river, tied his shoes to it and started to drift down, with the intention to swim ashore in a short time and walk back.

After floating awhile he perceived that he was being gradually borne away from the bank, and he then made an effort to disengage his shoes. Failing in this, because the strings had become wet, he at last tore them loose and started to swim ashore. But by this time he had got so far out from the shore that, chilled as he had become, he reached it with difficulty.

Naturally the inquiry arose, '' Why did the tree-lap drift away from the shore?'' and many a time in after years the question recurred. Wherever opportunity offered, streams were scrutinized with a view of finding an answer.

Step by step a little headway was made, until finally a few years' residence on the banks of the Mississippi supplied an opportunity for the completion of the theory.

THE LAWS OF RIVERFLOW.

WHEN we contemplate the surface of the earth with regard to its fitness as a dwelling-place for man, this appears to hold a vital relation to the behavior of the streams of water that exist upon its surface.

The water precipitated from the clouds at first collects into small bodies, each of which carves for itself a channel along which it flows, presenting an altogether pleasing alternation of pools and shallows. This arrangement renders all but the very smallest of streams fit homes for a teeming life which, while reveling in a joyous existence, yet serves as food for man.

Gathering next into rivers, the water seeks the great ocean reservoirs, whence, under the wooing of the sun, it may again return to revive the dry land, to sustain the various tribes of living forms to which it affords a dwelling-place, and then to carry back in turn a grateful tribute to feed the innumerable denizens of the deep.

How these rivulets, rivers, and seas that not only contribute so much of useful service to the inhabitants of the earth but have also chiseled every form of beauty presented by its surface have themselves been brought into existence, is a subject worthy of patient and diligent inquiry. It is under the conviction that new light may be shed upon the mechanism and laws which have been employed and

observed in their production that the author has
been led to make publication of the views embraced
in the present effort, firmly confident that the baf-
fling secret has been revealed.

THE BEGINNING OF THE SEAS.

When the incandescent mass of which the earth
at one time consisted first became sufficiently cooled
to allow the clouds of vapor which surrounded it to
condense and reach its surface in the form of water,
the spots where the water fell earliest and most
abundantly cooled off more rapidly than others.
The parts so cooled contracted and acquired thereby
a greater specific gravity than the surrounding and
hotter superficial masses ; and, being weighted down
by the waters which had accumulated upon them as
a result of this depression, they sank still further
into the liquid mass, as a partly filled bowl would
sink into a vessel of water.

Since the polar regions were the first part of the
earth to cool off, this beginning of the seas must
have taken place at the poles. As the cooling of
the earth progressed, the areas of sea formation
increased in extent, until it became possible for water
to remain on the entire surface of the earth.

Until a solid crust of some considerable thickness
was formed, there must have been, from a multitude
of causes, an almost continuous shifting of the surface level. Thus different rates of heat-radiation
from the earth's surface, deformation due to the
attraction of the moon and sun, and even to a slight
extent that of other nearby heavenly bodies, as well

as varying chemical combinations in the deeper parts, must all have resulted in more or less disturbance of the level.

POLAR UNDERCURRENTS.

After the sea had attained a considerable depth and the thickening crust had been divided off into ocean bed and dry land, cold undercurrents set in from the direction of the poles, and the earth at the bottom of the seas was cooled off still more rapidly than before, and more rapidly than the elevated or dry lands.

For, aside from the influence of this polar undercurrent, water is a better agent for extracting and dissipating heat than the atmosphere.

Another active agency in lowering the specific gravity of the dry land and elevating that of areas covered by the sea has probably been the continuous filtering of various salts from the elevated lands and their transference to the ocean bottom. For, as the bed of the ocean by its subsidence gradually forced upward and elevated the dry lands, water derived from rain and snow has percolated them everywhere and dissolved out vast quantities of soluble substances and carried them into the sea, both by way of the rivers and by that great subterranean movement of seepage that is continuously going on in the direction of the ocean.

This process has left the dry land more or less honeycombed throughout and of diminished specific gravity. On the other hand, of this extracted matter, the soda and magnesia salts alone have remained in solution in the sea water in any large proportion, nearly all the others having settled down into the rocks forming the bottom of the sea, and by adding to the density and weight of the sea bottom, by so much increased its subsidence. Thus the greater thickness and greater specific weight of the earth's crust forming the sea bottom causes it to overbalance the dry land and produce an equilibrium, notwithstanding the fact that it may be covered several miles deep with water whose specific gravity is only a little more than one fifth of that of the mass of the earth.

If all the causes in operation have been taken into account, it must therefore result that two cones of equal angles taken, the one from the sea and the other from the average level of the surface of the earth, or for that matter even from the highest mountain, both coming to a point at the center of the earth, would be of equal weight.

It is certain, however, that there are other but unknown forces at work determining the alternate rising and falling of the crust of the earth, or at least of its surface. All the known forces make for permanence ; all tend to deepen the sea and to elevate the land. And yet the part of the earth's surface that now constitutes the dry land has been submerged beneath the ocean waters and again has risen above them times innumerable. How this has come about no one professes to know, and not even a guess has been ventured in explanation.

Though scarcely relevant to our discussion, the vast elevations known as mountains may be mentioned as offering like difficulty. Mathematicians have sought to calculate the strength of the earth's crust that would be adequate to the sustaining of the mountain ranges and preventing them from settling down to the earth's level. Yet, who knows that they bear heavier on the earth's surface than does an equal area of the plain or of the ocean? It may be that they actually superimpose an additional weight upon that part of the earth where they are situated, and that they now rest upon a rigid support that prevents them from sinking down with the part of the earth's crust that is beneath them.

Still, they must have been lifted up while yet resting on a plastic foundation, otherwise they could not have been elevated with their present contour. Why did they not, then, subside while their foundation was yet plastic, or who is able to say that they are not now supported by the same force that raised them? The same mighty power is doubtless at play with the mountains as with the ocean bed, though its methods are as yet unknown.

The operation of this mechanism, viz., the subsidence of the ocean and the elevation of the land, is probably the chief factor in the production of volcanoes, and largely a factor in the production of earthquakes. If a piece of cardboard and one of writing-paper be pressed against each other edgewise, the thin paper will bend before the cardboard, and if they rest on an unyielding surface the convexity of the bend in the thin paper will be upward and close to the edge. In like manner the thin dry-land crust, when pressed edgewise against the stronger sea-bottom crust by the force of the earth's shrinkage, bends with an upward convexity near the sea in such a way as that the curve near the edge of the sea bottom, as well as that near the edge of the land crust, constantly becomes sharper.

Every increase in the curve of the sea-bottom crust will produce a number of ∧-shaped fissures in its under surface ; and every increase in the convexity of the adjacent land crust will produce a number of ∨-shaped fissures in its upper part. Now, into the ∧-shaped fissures, when formed, expansible material will escape, and often large masses of sea water. when these fissures happen to extend up through the sea-bottom crust, will drop into them and be exposed to intense heat.

The mass thus suddenly heated and enormously expanded can not easily escape through the apex of the ∧-shaped fissure, for the walls of its apex will consist of solid material, held together arch-like by the enormous weight of the sloping seashore above, capped, as it often is, by great mountain ranges.

The expanding mass will therefore pass up under the land crust, where, finding the ∨-shaped fissures extending down to the softened incandescent mass beneath, it will escape through it, and thus give rise to a volcano. Or, failing to find such a vent, it will spread out as a wave and give rise to an earthquake.

It is well to note that even if the sea-bottom crust is no thicker or heavier than the land crust, but only equally as thick, it must, other things being equal, be made more rigid by reason of the weight of the ocean water resting upon it.

Nor is it quite impossible, judging from what we now know of the vaporization of metals and their movement under electrical influence, that the successive rising and subsidence of the earth's surface are largely due to shifting of metallic masses determined by the power of the great earth-currents of electricity.

THE GENESIS OF RIVERS.

As soon as the level of the sea became sufficiently lowered, drainage began from the elevated lands and initiated the formation of streams, a part of which ultimately became rivers.

These grew with the widening lands through the long æons of time ; and when man appeared upon the scene, they must, from the earliest unfoldings of intelligent curiosity, have been among the chief features that attached him to his dwelling-place. They yielded at first, most probably, his main supply of animal food. They formed one of the earliest and often one of the easiest means of intercommunication, as well as the natural boundaries of countries. And with the birth of civilization the waters of the rivers made the desert bloom and yield the readiest of all attainable harvests ; and thus they became in many lands objects even of divine adoration.

They must have been regarded from the beginning, as they still are, as among the most pleasing ornaments of the landscape. Nor have they yet ceased to supply the painter with an exhaustless theme and the poet with a generous inspiration.

Sustaining so many important relations to the happiness and well-being of mankind, it is reason-

able to suppose that they were from early ages
objects of curious regard, as they have been in more
recent times of much profound research.

And yet of the various important problems that
have arisen from time to time, having a bearing
upon the true theory of streams, many remain to-
day confessedly without solution, and many striking
phenomena wholly unexplained. Included in this
category, the following questions present themselves as entirely unanswered, and must be regarded
as the touchstone or criterion by which the truth of
any theory is to be tried; for, with the facts they
indicate, any true and consistent theory must accord :

(1) Why are there brooks and rivers? By what
means has water been enabled to cut channels
through masses of solid substances having a specific
gravity two and a half to five times its own?

(2) Why are the channels of streams trough-
shaped, or representations of a segment of a cylindrical tube?

(3) Why are there deeps and shallows in streams,
and why do rivers enter the sea over beds sloping
upward?

(4) Why does floating material drift from the
margins to the middle of streams?

(5) Why is the swiftest point, or *locus* of greatest speed, in streams not at the surface, but at a
considerable distance beneath?

(6) Why is the surface of streams not level when
measured on cross-section, but higher in the middle than at the margins?

(7) Why do rivers flowing through deltas throw up elevations or natural levees along their banks?

(8) Why do rivers entering the sea through deltas of their own forming have multiple mouths?

(9) Why does water moving in steep channels attain so much less speed than would solid bodies under like conditions?

The correct answer to all these questions the writer believes to lie in a correct interpretation of a principle underlying the motion of all liquids as well as fluids, which he believes he has discovered, and which he has denominated " *The law of the double spiral.*"

(I) WHY THERE ARE BROOKS AND RIVERS, AND HOW IT IS THAT WATER IN STREAMS HAS ERODED CHANNELS IN SUBSTANCES HAVING A SPECIFIC GRAVITY GREATER THAN ITS OWN.

To illustrate the principle, let us begin with a stream in the first steps of channel formation. Let us suppose a quantity of water to be steadily poured upon a smooth surface of erosible material, and one of sufficient incline to determine motion of the liquid; and let us farther conceive this stream of water to consist of columns of molecules.

At first the water will move down the incline in a thin stratum, limited and restrained on either side by a wall held together by that form of adhesion known as surface tension; just as a drop of water on the floor will remain inclosed in a wall of its own particles and retain its form.

In the stream thus supposed to be formed the column of particles at the outer edge on each side and next within these walls will be retarded more than any of the columns within; and as friction at the bottom must be greater than at the top, this will cause the particle at the lower end of each external column to be most retarded of all the particles. Now, not only will this lower most outward molecule be retarded, but it will be progressively and continuously retarded; so that if a new impulse is not given it, if new force is not imparted to it, it must in time come to an absolute standstill.

However, the lines of particles following each other in this order, that is, nearest the bottom and nearest the edge, will not be uniformly retarded even in the smoothest channel; for the lines will be pulled apart times innumerable. And since the width of the stream must be maintained and its area conserved in order that a constant quantity of water may pass any given point, whenever such a break begins to occur, that is, whenever a gap in the line is forming, it must be filled from the column of molecules next within, for the reason that there is none on the outside from which it can be filled.

Now, the column of particles next on the inside has a greater speed than the outer one, and from whatever part of such column the particle may be supplied, it will be moving faster than the one it may have supplanted.

If a rod were thrust to the bottom of a mass of water, however deep, and then drawn out instantaneously, the hole thus made would fill first at the bottom from the contiguous particles surrounding it. Likewise where there is a tendency to form a vacuum, and a suction is produced thereby, this forming vacuum will be supplied or filled by a molecule from the bottom of the next column within.

As this other molecule moves out, its place must in turn be taken by some other. The one before it is moving on ; the one behind it would have to increase its speed, but is itself being retarded ; so this vacuum, or tendency to a vacuum, must be met or supplied by the next molecule within. This process will continue until the middle of the stream is reached from either side.

When the stream has been equally divided on the basis of retarding forces and the middle has been reached from each side by this outward movement, the molecules displaced either by further retardation or by the outward movement below must be replaced from above. This will cause a constant downward movement at the middle of the stream or in the line of the current.

The outward movement of the particles at the bottom of the stream will be attended by a momentum proportionate to its extent, and the result will be that as these outward moving particles or masses strike against the limiting wall or bank, and can go no farther, the water there will be lifted up and an elevation or ridge will be formed along each edge of the stream, so that its surface will take the form of a trough.

Having now in the smallest supposed stream illustrated the molecular movement, we will next take a larger stream for the purpose of easier understanding, and suppose it to have reached the stage where the water has been heaped up at the sides by the outward movement below. The surface of the stream will now form a longitudinal depression or trough. Each half of this trough will have at the surface an incline in two directions: one toward the depressed middle of the stream and the other downward in the line of the stream bed or channel.

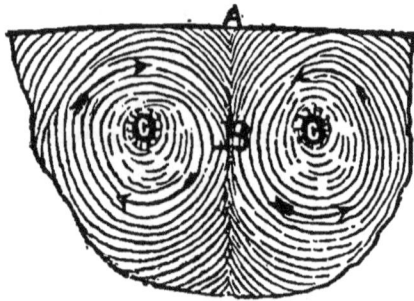

TRANSVERSE SECTION OF STREAM.

A—Elevated middle of stream.
B—*Locus* of greatest speed.
C—Center of cylinder.
The arrows show the direction of the spiral motion.

The water which has been heaped up at the banks by the momentum of the outward undercurrent will, therefore, as a resultant, flow obliquely downward and inward, by a countercurrent at the surface, from the margins to the middle.

At the same time, the water having been drawn from the bottom at the middle of the stream, toward each of the banks, the overlying portions will sink down to take its place. In obedience to these forces, therefore, every stream moving in a channel formed of resisting material of necessity resolves itself into two equal cylinders revolving spirally on parallel axes in opposite directions ; that is, *outward at the bottom, upward at the margins, inward at the top, and downward through the middle.*

We may now pass to a consideration of the validity of the response which this principle, denominated the law of the double spiral, offers to the test questions already suggested as a criterion.

If, as generally held, water moves along in streams by strata or layers, one gliding over another, the result of such action could not be the formation of channels. In such case, scour could be effected only by the bottom layers, and these after a short time would become laden to the limit with silt. Since the mountain streams would take up and bear along with them more of silt than the lower layers in the large streams could carry, the tendency would be to drop it on reaching the valleys ; and the effect would be to fill up rather than to scour out channels.

Nor is there any reason to believe that on any principle hitherto recognized the action of water would tend to the formation of channels. No reason can be given in the present state of science why water should not spread out over the land indifferently, and thus make its way by a slow, creeping movement to the sea, instead of carving out channels and traveling in them.

The explanation ready in the minds of most people is, that rivers are formed by the water seeking the lowest continued depressions along constructional slopes. But this explanation, at first blush apparently so obvious, fails when we examine the rivers of deltas, or rivers that have reached the sea through deposits of silt with which they have encroached upon the ocean. In these situations it will commonly be found that the rivers have actually carried detritus out into the sea and walled off the sea water by building up banks for themselves, and thus made their own channels.

At the mouths of the Mississippi one may travel for miles on boats through the passes so near to the sea water, walled off by the action of the river, that a stone may be thrown into the Gulf across the banks that reach out like so many fingers into the sea.

(2) WHY THE CHANNELS OF STREAMS ARE TROUGH-SHAPED OR FIGURED AFTER A SEGMENT OF A CIRCULAR TUBE.

It is obvious that when the water of the upper part of a stream leaves the banks to begin its flow to the middle, it continues to increase its speed, not only until it reaches the point where in turn it must change its course obliquely downward toward the bottom, but also for some distance beneath the surface.

The water of the most rapid part of the stream, therefore, has also the most direct course downward against the bottom, and so in the line of this part, which is the current, the greatest extent of scour or erosion must take place. Every stream, then, other things being equal, must be deepest in the middle or in the line of its current. The water passing down from the surface increases in speed until it traverses about three tenths of the distance to the bottom, when it begins to suffer the retardation resulting from the resistance of the bottom of the channel, so that when it reaches the bottom its speed is nearly the same as, or even less than, that at the surface.

This retardation now steadily and progressively increases as the water at the bottom moves obliquely outward toward the bank, and the erosion of the bottom of the channel steadily decreases with the retardation of the water. At length the speed becomes too small for any erosion to take place, when the water, rising toward the top, begins again its movement to the middle. This movement for any stream must determine its limit as to width, and result in the formation of a trough-shaped channel.

There is probably also a far wider application for the principle than the one here under consideration.

As will be shown farther on, water in streams moves in a most irregular way, little masses being projected inward and outward, upward and downward, continually. As long as the irregular masses come in contact only with other portions of the stream, they expend their momentum in producing widespread and uniform disturbances of equilibrium. But if they meet with resistance from a solid, the impact produces increase of pressure at the solid surface.

It is evident that this increase of pressure in a stream, being directed mainly against the stream walls, is an additional factor in determining an outward and upward movement of the water along the sloping bottom and sides, respectively.

It is true that, strictly speaking, no actually trough-shaped channel exists ; yet the form of every channel eroded through homogeneous materials is in the mean that of a trough.   For if one of the banks is abrupt, the opposite one will have a gentler slope than usual, and this will compensate for the depart- ure of the other from the natural order or average.

But there must also be a law of limit as to the depth as well as for width ; otherwise streams would go on indefinitely deepening.

Whatever the slope of its bed, every stream consists of a succession of deeps and shallows, but these always bear a relation to the size of the stream ; and no matter how soft the material of the bed may be, the general depth is proportionate to the size of the stream.

The reason given by Paul Frisi, one of the most eminent of the Italian engineers, and the cause gen- erally accepted in his time for the arrest of the deepening of channels and pools was, that after reaching an uncertain depth the bottoms of stream beds become paved with coarse sand and gravel where they do not already consist of rock, and that thus further erosion is prevented.   And curiously enough, up to this date, so far as the writer knows, no other explanation has ever been offered.

Every one, however, who has made any practical study of streams knows that this is not the fact. Gravel and sand in creeks, and boulders in rivers, are not found in the deep places but in the shallow places, that is, on the riffles. The cause of the arrest of erosion must then be sought elsewhere.

Sir Charles Lyell gave it out as the result of a number of experiments made under his supervision, that water moving at a velocity of forty feet per minute will sweep along coarse sand; one of sixty feet, fine gravel; one of one hundred and twenty feet, rounded pebbles; and one of one hundred and eighty feet (a little more than two miles an hour), angular stones of the size of an egg. These experiments have been accepted as authoritative by many writers on the subject, and appear in several encyclopedias.

What the depth of the water was, employed by Lyell in his experiments, I have not been able to ascertain. Evidently they are not complete, as leaving out of the account larger masses of rock that we know are moved by the force of streams; and it can probably be shown, also, that they are in other respects misleading.

To reach the conclusion required by the explanation about to be offered for the phenomenon in question, it becomes necessary to modify not only the results given out by Sir Charles Lyell, but also another misapprehension some rest under in the matter of hydraulics. It is held that motion of a body in water is not affected by the amount of pressure on the water, though this really applies only to the motion of bodies without appreciable thickness. Otherwise this would be to claim that a fish with the same effort might swim as rapidly in water a mile deep as a fathom deep. If this were

to hold without modification, the depth of the water
would count for nothing in the matter of lifting and
transporting any form of detritus.   After the depth
of water became sufficient to cover the object to be
moved, its removal would be a question of the speed
of the water alone.

But we have just seen that gravel, sand, and
boulders are found not in the deeps in creeks and
rivers, but in shallows or on riffles ; so that, if a
boulder be cast into a deep place, it will be picked
up in times of flood and carried to the next shallow
below, and there dropped into water swifter than
that from which it was removed.

This will often happen with stones of large size.
From a ledge of rock just below Brandenburg, in
Kentucky, as the writer has observed, large masses
of limestone have been torn away by the Ohio
floods, and, after being carried some three miles
further down, cast out onto the overflowed banks.

The erosive power of a stream must then have a
relation to the depth as well as the speed.   This
may not be and probably is not directly as the
product of the speed by the depth, but it is doubtless some fixed product of the speed by the depth.
As the relation is most probably constant, whatever
it may be, we will for purposes of argument assume
that it is direct.

To make this better understood, we may conceive a stream to occupy a rectangular canal.   Now,
it is easy to perceive that if the water is made just
half the depth, it must move just twice as fast, in
order that the same quantity shall pass a given

point in a given time. If we take away one third the depth we must add one third to the speed, and so on reciprocally for all other proportions.

Assuming, now, that erosive power is the product of the speed by the depth or the weight of water, we obtain a numerical formula that bears out observation.

In a canal such as we have supposed with a constant width, let a be the normal depth and b the normal speed. Now, any variation of a and b from this normal giving the same sum will give a diminished product, whether that be the increase of a and the lessening of b, or an addition to b by a subtraction from a — that is, a multiplied by b will give a larger product than any other numbers into which the sum of a and b can be divided.

If a is four and b four, their product will be greater than that of any other numbers into which their sum, which is eight, can be divided.

This can readily be gathered from the annexed table, thus :

Depth.	Speed.	Sum.		Depth.	Speed.	Product.
8 +	0 =	8		8 ×	0 =	0
7 +	1 =	8		7 ×	1 =	7
6 +	2 =	8		6 ×	2 =	12
5 +	3 =	8		5 ×	3 =	15
4 +	4 =	8		4 ×	4 =	16
3 +	5 =	8		3 ×	5 =	15
2 +	6 =	8		2 ×	6 =	12
1 +	7 =	8		1 ×	7 =	7
0 +	8 =	8		0 ×	8 =	0

We find, then, that if a stream varies in either direction from a certain relation of speed and depth, which may be denominated the normal, erosive power diminishes, finally ceasing altogether as zero is reached on the one hand or the other.

Not only does the general depth of streams depend upon the conditions expressed in the foregoing formula, but also the production of the deep pools that in nearly all streams alternate with shallows. However, the positions these formations occupy relatively to each other must be due to some other modifying influence.

Since we find deep places far apart in the larger rivers and close together in the smaller streams, we would naturally infer wave action as the modifying force. And so no doubt it is to a very large extent, but these deeps and shallows are so irregular in frequency and length that we can not avoid suspecting that some other force is in operation besides wave action. In another essay of the series, which relates to the action of the liquid wedge, we may find an auxiliary force that will supply the needed causative influence.

But whatever it may be that determines the production of deeps and shallows, in addition to accounting for the arrest of deepening in these places, we must account for the fact already mentioned, that in deep places masses of heavy material such as boulders are picked up by the water during floods and dropped in the faster moving shallow water below.

This, too, will be found embraced by the formula. To account for it we assume that during floods the same additional height is added to the deep and the shallow parts of the stream. This adds more to the speed of the deep water than it does depth to the previously shallow water, and so sends the normal speed-point downward faster in the deep water than in the shallow, and the product of the depth by the speed now becomes greater in the deep water than in the shallow.

It is not to be lost sight of that in the short pools as in the long river stretches the acting forces are subject to the regimen established by the next obstruction below, which, if removed, would allow the channel of the affected part to be cut deeper.

(3) WHY CHANNELS OF STREAMS ARE AS A RULE DEEP NEAR THEIR ENTRANCE INTO THE SEA, AND YET ENTER IT WITH BOTTOMS SLOPING UPWARD.

At the mouth of nearly every stream entering the sea there is a shallow place in the channel near the entrance, but immediately above this shallow or bar the water is almost invariably deep. The rule is, therefore, that rivers enter the ocean over channel beds sloping upward. The cause of this bar at the mouth is simply the precipitation of silt from the river water, due to the arrest of its speed on reaching the sea. But the deepening of the channel above this bar has a different cause.

When the tide is flowing in at the mouth of a river the water is dammed back until it is level with the adjacent tide. As the tide falls the ocean water can retreat more quickly than the river water can flow out, and the result is a great slope of the water in the river where it joins the sea.

At first, only the water above the sea level will escape, the lower part of the river being undisturbed. This soon brings about a condition very favorable to the development of the double spiral; for a head in the river a little way up, caused by the outflow of the upper part of the water next to the sea, will quickly start the water to rising at the edges, and will thus inaugurate a brisk flow, both spirally and longitudinally. The result will necessarily be the boring out of a deep place in the river channel near the sea. It is a question whether there is not a tendency to bore out deep places in streams above all obstructions, partly due to shock and partly to the causes just considered.

(4) WHY FLOATING MATERIAL DRIFTS TO THE MIDDLE OF STREAMS.

As a result of the double-spiral movement of water in streams, and one of the strongest proofs of its existence, may be cited the well-known fact that floating material tends to drift from the margins to the middle of all enchanneled streams. This is unvaryingly the case.

It is true that when rivers are low, as in summer, and the current is very slow, if the banks are high the attraction of gravity will cause floatage to move to the shore. Winds at any time may carry drift to the shore. All this, however, does not alter the tendency of floating material to move toward the middle of streams, but merely overcomes it for the time being.

Now, no one will insist that this drift skips across the water. The water that carries it to the middle must go with it. If, then, it be true that the water at the surface of a stream, everywhere and perpetually, moves from both sides toward the middle, *the conclusion is absolutely irresistible that it must sink down in the middle and return to the margins by a countercurrent beneath.*

At the meeting of the Mississippi River Commission at St. Louis, in 1884, Gen. John Newton, Chief of Engineers, U. S. A., kindly submitted this theory to the commission with his approval, a copy having been sent him for consideration by Gen. W. B. Hazen, Chief of the U. S. Signal Service.

Objections were offered against it at the meeting on the ground that pilots and other river men had observed that floating material is carried away from the middle of streams in rising water, and that it is more difficult to keep tows, rafts, etc., out of the banks in rising water than in falling, and that it is only in falling water that floating material drifts to the middle.

The matter was again brought forward by Gen. Newton at the next succeeding meeting of the American Society of Civil Engineers, at Brooklyn, where it was met with the same objections.

The first of these objections, if valid, would be fatal to the theory; and it is strange with what uniformity it is urged by men whose life is on the river. It is in no sense true, however, being based on an error of observation that is self-evident, while the phenomena embraced in the second objection do not oppose but bear out the theory.

When a pilot is directing his boat on a river he
is looking only for a clear path, and not seeking to
ascertain in what direction each piece of drift is floating.   He thus fails to take note of the fact that the
drift is spread over the river, not because it is moving away from the middle, but because it is moving
from various points at the bank obliquely toward
the middle, which it has not as yet had time to
reach.

To realize the impossibility of the first movement
of drift being from the middle to the margins of
streams it is only needed to reflect that in low
water there is no drift, either in the middle or any-
where else in a stream.   How, then, seeing that a
river must rise before it falls, can the drift move first
from the middle to the banks ?   It must certainly
be borne to the middle before it can be carried from
it.   As to the increased difficulty of keeping river
craft out of the banks in rising water, that fact is
perfectly consistent with the theory.

In a straight stream the two spirals lie side by
side, with the water descending in the middle.   But
on reaching a sharp curve the inner spiral is tilted
over the outer one, and thus instead of two spirals
side by side, the stream will in this case consist of
an upper and an under spiral, with the part of the
water constituting the current sinking down at the
outer bank instead of the middle.

As the current will in this case also be at the
outer bank, the boat crowding toward it will be
simply drifting into or with the current, in accordance with the invariable rule.

On the contrary, at the bottom the movement of sand and pebbles, and in large, rapid streams of rocks of considerable size, will be obliquely outward. On examining the bed of a stream of ordinary dimensions after a flood it will be observed that the coarser gravel occupies the middle of the stream ; on the outside of this the finer gravel will be seen ; next the coarser sand, then the finer sand, and finally a deposit of mud or clay.

These deposits will often be found in the form of banklets pointing obliquely upward and outward from the axis of the stream, and thus furnishing both proof and illustration of the fact that the water at the bottom had flowed obliquely outward and had been progressively retarded. Clearly, only the heavier gravel could be dropped where the water was swift, and after that, the finer particles in turn as the water lost its speed.

(5) WHY THE POINT OF GREATEST SPEED IN STREAMS IS NOT AT THE SURFACE.

For a long period, doubtless from prehistoric times, it has been known that the water in a river or other stream is swifter some distance beneath than immediately at the surface. Men who were much on rivers often observed that if a skiff was loosed from a boat, the boat drifted ahead of it, and if an oar was dropped from a skiff, it lagged behind. It was easy and early perceived that the cause of this was, that the boat reached down into a faster moving stratum of water than that which carried the skiff, and that the oar, which floated lightest, must be moving in water that was the slowest.

The first and for a long time the only explanation of this phenomenon was that the mere friction of the atmosphere against the surface of the water was sufficient to cause the observed retardation.

This was the view with which Humphrey and Abbott closed their elaborate investigations made in 1846 in connection with the government survey of the Mississippi. They found, as a result of numerous measurements, that the point of maximum speed for the Mississippi was about three tenths of the depth of the stream beneath the surface. Boileau found the maximum velocity, though raised a little for calm weather, still at a considerable distance below the surface, even when the wind was blowing down stream with a velocity greater than that of the stream, and when the action of the air must have been an accelerating and not a retarding action. He found also, as did Humphrey and Abbott, that the depth of the maximum speed varied with the wind. When the wind blew up stream the depth was greatest, and it was smallest when the wind blew down stream.

Prof. James Thomson has given an explanation of the diminution of the velocity at and near the surface, or, rather, of the increased speed beneath the surface, which the writer of the article on Hydrodynamics in the Encyclopedia Britannica pronounces much more probable than the theory of friction against the atmosphere.

He suggests that portions of water, with their velocity diminished from retardation by the sides and bottom, are thrown off in eddying masses and mingle with the rest of the stream. These eddying masses modify the velocity in all parts of the stream, but have their greatest influence at the surface. Reaching the free surface, they spread out and remain there, mingling with the water at that level and diminishing the velocity which would otherwise be found there.

This rising to the surface of the water that has been retarded by friction against the sides and bottom of streams is exactly what the theory of the double spiral requires, but not at all in the way that Professor Thomson suggests.

In the first place, neither pressure on the water constituting the lower layers of the stream nor its retardation would operate to carry it to the surface, either in eddying masses or otherwise.

In the second place, if masses of water should begin to rise to the surface after or because their motion had been retarded, it would be necessary for them to make their way through the more rapidly moving central portions, and they would necessarily acquire thereby a similar motion.

This suggestion of Professor Thomson is, however, a most important one, as coming from an acute observer and profound thinker, pointing out, as it does, the only possible source of retardation.

It only needs to supplement it with the assumption of a systematic method by which the retarded masses regularly and continuously spread out over the surface.

On the principle of the double spiral the explanation becomes simple and easy. The strata of

water in contact with the channel walls are retarded
by friction as they steadily make their way along
the bottom toward the edges of the stream, this
retardation diminishing from below upward. It, of
course, reaches its greatest extent at the margins of
the stream, and the water there rising, spreads out
over the surface, or rather rolls over toward the
middle as the upper half of the stream.

It is not to be supposed, however, that each
spiral revolves as if made up of so many layers like
sheets of paper that steadily keep their places. All
along the edges of streams, and in fact all over the
bottom, there will be irregular breaks. The water
will fly off in diminutive masses here and there, and
near the edges of the streams these will be seen
boiling up through the free surface. A beautiful
illustration of this may at any time be observed by
looking down from a bridge onto the surface of a
swollen stream.

Probably a fourth of the stream on either side
will be seen boiling up irregularly throughout its
whole extent ; within this will be seen a smooth belt,
while at the middle there appears a narrow band
rough with waves or ripples.

This rough surface is probably due to slack as
well as to greater speed, the water moving in from
the sides faster than it can sink down in the middle.
The roughness of the small band in the middle of
streams, known as the current, can not result from
mere speed alone, because the smooth band on each
side of the current may at the same time be moving
very much faster in such a stream than the rough-
surfaced current of the smaller streams.

The position as well as the origin of the *locus* or transverse line of greatest speed is likewise susceptible of explanation on the theory of the double spiral.

Ordinarily the most rapid part of a stream, as already stated, is three tenths of the distance from the surface to the bottom. This would allow one half of the stream to be above and the other half beneath the swiftest part, for while the upper half is the wider and the more rapid, the lower half, though the slower, is the deeper. When the wind blows up stream, the swiftest point goes deeper; for in that case the upper part being made to move more slowly, it must have a larger volume in order to form half of the movement. On the other hand, when the wind blows down stream, the upper half is more swift, and does not need a depth of three tenths in order to embrace half of the movement of the water in the stream, and therefore the maximum of speed rises.

The controlling condition obviously is that the same quantity of water must move out below the maximum line that moves inward above it.

Reflection will show, also, that the half of the river that is moving in above, and the half which is moving out below, must have uniform velocities at all their points of contact. Thus the upper half is continuously gaining in speed as it approaches the middle, the highest speed of both being at this common turning point in the middle. Returning to the bank, then, the under half, which was the upper as it went in, will lose speed at exactly the same rate

it gained it on going in, and so from the top quite
to the bottom there will be at all points a relative
equality, or rather a regular gradation of speed.

(6) WHY STREAMS ARE HIGHER IN THE MIDDLE THAN AT THE MARGINS.

The surface of streams has been described as
trough-shaped at the time of the beginning of the
inward flow of water at the stream surface. Yet,
as an actual fact, the stream surface probably never
presents this form unless it may be just below dams
or cataracts. On the contrary, the middle is probably always as high as the margins, and in rapid
streams really higher.

At the Yukon rapids the middle of the river is
said to be six feet higher than the edges. This fact
of the elevation of the middle of rivers has been
long observed, and two different hypotheses have
been offered in explanation : one by Sir Isaac Newton and the other by Major Allan Cunningham, R.
E. of the English Army in India.

The explanation offered by Major Cunningham
is that pressure being diminished by speed, and the
speed of those parts of a stream forming the current being the greatest, the water in the line of the
current becomes elevated.

This notion is probably derived from the movement of gases through pipes or tubes, and most
likely does not in the least apply to the movement
of water in open channels. When gases are forced
through pipes, they of course get relief from pressure by escaping at the distal end with increased

speed, but it is not easy to see how this can apply to an incompressible liquid like water moving in an open channel.

If the rising up of the water is a function of its speed depending on diminished pressure in these cases, then it is hard to see how it is that the water of very swift streams, such as that of the Yukon rapids, does not leave its bed and fly off into the air.

The explanation of Sir Isaac Newton is not materially different. He held that friction at the margins of streams pulls down the water, while the middle, being less subject to friction, obeys its momentum and remains elevated in the form of a ridge.

To this there are many objections. In the first place, no such momentum could pertain to water as would be required to hold the middle of a stream up in this way. Let us suppose that a stream, with a fall of four feet to the mile, moves six miles an hour and has the middle elevated six inches. This means that the water of the ridge would fall four feet in an hour, an almost level projectory for a distance of six miles and lasting one hour; for we must assume, according to Newton's theory, that the middle stays up by its momentum. Now, the most powerful cannon known does not have a projectory so nearly level for five seconds. Again, say a stream moves so rapidly that after falling over a dam and losing its ridge, at the end of a mile it has regained it, and say that its ridge is four inches in height. At two miles it ought to be eight inches, at four it should be sixteen, and so on until the stream should stand on edge like a board.

Still another objection is the conclusion that if the friction of the banks pulls down the edges of the stream, the friction at the bottom of the bed ought also to pull down the bottom of the stream in the middle. But as the top remains in the air on account of its momentum, this should result in making all rapid streams hollow in their center.

When the water that forms the sides of the trough which every stream is resolved into by the momentum of the outward undercurrent begins at any point its flow obliquely back to the middle, the result is the filling up of the part of the trough just below. As the water approaches the middle on either side, the friction it has to overcome grows less and less and its speed greater and greater, until it reaches the water coming from the opposite side.

The tendency of the action of gravity of course would be to reduce the surface to a level, but the momentum of the water is sufficient to overbalance the indirect influence of gravitation and raise the middle of the stream above the general level.

The concavity of the stream surface at the cross-section will, therefore, not only be filled up by the water which has flowed obliquely inward from the banks above, but the middle will even be raised by it into a ridge.

Thus every stream presents the paradox of a surface that is at the same time a trough and a ridge ; potentially a trough and actually a ridge. Furthermore, if a river could be made to cease its motion and remain still until its surface became level, and then allowed to resume its wonted motion, the water of the edges would at once rise, and the surface would become an actual trough. Ordinarily, however, only the ridge is actual on the transverse measurement. But if across the surface of a stream we measure a curve with the convexity downward in the direction of the flow, we will invaribly have an actual trough. That is, while the middle at any point on transverse section may be higher than the margins, the particles forming the raised middle have not flowed there from a lower level. On the contrary, each particle of the water forming the ridge will be found on a lower level than it occupied when it left the bank above. No particle has flowed up hill.

Fig. 2.

SECTION OF DELTA RIVER BED.

(7) WHY DELTA STREAMS THROW UP NATURAL LEVEES.

All rivers reaching the sea through deltas made of deposits of their own silt have their banks elevated so that they traverse channels apparently scooped out of the crest of a low ridge.

A cross-section of a delta river will show that the highest part of the land is that immediately next to the river, while the surface on each side gradually slopes outward in both directions from the main channel. The only arable lands along the lower parts of many rivers consist of the elevations formed of silt which has been cast out upon the banks either by the main rivers or the innumerable outlets they have had from time to time as they pushed their way farther and farther into the sea. As already pointed out, the regimen of the lower sections of all rivers is controlled by the sea.

During floods the transverse currents at the bottom of a river throw the silt out over the banks, where, on account of the slowing of the flow, it is deposited. Now, the part of the bank nearest the river always consists of the coarsest particles of which the silt is composed. Outside of this coarse matter finer silt is deposited, and still farther on the very finest particles, often so fine, indeed, that they are easily spread out, covering large areas and forming extensive plateaus almost as level as the sea.

Year by year and century by century the river gains in length by deposit of silt in the sea near its mouth, and with each increase in length a higher level is required for the stream bed all along its course, higher banks are needed to restrain overflow, and these grow by accretions of upheaved and ejected silt.

(8) WHY DELTA RIVERS HAVE MULTIPLE MOUTHS.

Rivers that have formed deltas at their mouths have without exception multiple outlets. The number of these outlets runs sometimes into hundreds, and, as in the case of the Ganges, possibly even into thousands.

When a river reaches the sea the speed of its water is arrested by reason of its spreading out over the heavier salt water, and a deposit of silt will take place, forming a bar at the point of entrance.

The spiral, as already explained, requires that there shall be a certain proportion between the width of the flow and the depth. When the width is too great the stream breaks up into a number of spirals, each one of them with its fellow beginning to cut out a new channel. Now, during the high water the stream might maintain a single double spiral over the bar, but as the water falls it becomes too shallow to return to the line of the current when it goes out, and it therefore breaks up into two or more or even many double spirals, each one of which begins to cut a channel that may in time become a new outlet. Succeeding floods will fail to entirely obliterate these notches in the bar, while each subsequent season of low water and the deposits of successive floods will lengthen them out into the sea.

As the incline of each one of these outlets must be extremely small, from the fact that this channel formation first began at sea level, the current will be slow, and it will become slower in proportion as the length of the outlet becomes greater. In the course of time the current in a part of these outlets becomes too slow to wash out the silt which is thrown into their channels at the source by the transverse undercurrent of the parent stream ; whereupon such outlets are filled up with silt at this point, and thus shut off from the main channel. In this way originate the numerous blind bayous that are found so abundantly in all deltas.

In the mean time, as the sea is caused to recede by reason of the deposit of silt, the main stream will be less interfered with in its spiral action. By the elevation of its bed, also, it will be still further removed from interference by the sea, so that in the course of time it will select that one of its mouths which proves to be the line of least resistance, and, closing up the rest where they break off from it, will continue to advance as one body further on through the delta.

(9) WHY WATER FLOWING IN STEEP CHANNELS DOES NOT INCREASE IN SPEED AS DO SOLID BODIES DESCENDING WITH THE SAME INCLINE.

When a solid body is caused to fall through space it increases in speed indefinitely, and the distance it will fall in any given time is the distance it will fall in one second multiplied by the square of the number of seconds it is falling. If permitted to slide down an inclined trough there might be some lessening of acceleration, but the rule would fairly well apply to the conditions if the opposing surfaces were uniformly smooth. With water, however, it is far different ; for in a very short time it reaches its highest attainable speed.

EXPERIMENTS OF NAPOLEON'S ENGINEERS.

When Napoleon, desirous of avoiding interruption at the hands of the English, sought to have ships constructed in Lake Constance, his engineers built a flume of logs framed together, in which to slide the timbers from the top of Mt. Pilatus down into the lake, a distance of some six miles.

At first they attempted to slide the timbers down the flume without water. So great, however, was the speed acquired by the timbers in their descent, that such heat was generated as threatened to speedily destroy the flume, while the timbers themselves were materially damaged by it. They thereupon directed into the flume a small stream of water which they found near its head. When, after this, the logs were let slide, they went down with such speed that when they reached the lake they plunged through many feet of water with sufficient force often to stick in the mud beneath. On the other hand, the stream of water increased in speed only for a few hundred feet, and then continued on to the lake without further acceleration. This peculiarity of motion which is found in all streams is known as kinetic equilibrium, and as to them has never yet been explained. On the principle of the double spiral, however, it admits of easy and satisfactory explanation.

HELIX RISES WITH THE SPEED.

Since the transverse current in a stream depends altogether upon the friction against its channel wall, it will of course increase its angle to the stream axis in proportion to the shallowness of the water and the rapidity of the flow ; that is, the helix of the spiral rises as the speed of the stream increases.

In a deep, slow-moving stream the spiral flow or helix will be very oblique, but in a swift stream it will be correspondingly more transverse and at a greater angle to the stream axis.

It results, therefore, that each particle of water in a mountain torrent flows a much greater distance in accomplishing a mile of progress in its channel than a like particle in a deep, slow-moving stream. The mountain stream dissipates the energy it gains by its fall in simply beating transversely against its banks. It is for this reason that the speed of streams has not hitherto been found to bear any constant relation to the incline of their channels. Thus the Rhone and the Amazon, where they flow through level lands, and have seemingly no fall, are nearly as rapid as the swiftest mountain torrent.

DETERMINATION OF THE SIZE OF CHANNELS.

It is the law governing the relation of width to depth that determines for each particular stream the size of its channel. Streams flowing from springs always carve out channels of appropriate dimensions and never overflow their banks; and the same may be said of all streams kept up by a steady flow of water.

If the Amazon, at its highest flood, could be turned into the Mississippi and the stream thus formed be kept at an even height, in the course of time it would regulate its channel and remain within its banks. All rivers would regulate their channels in the course of time, even in the season of floods, except for the continual lengthening they receive at their mouths and the excess of coarse detritus carried into their channels during floods.

So long as additions to the length of rivers are made by deposits of silt at their mouths, the banks along their course above, as already indicated, must be raised in order to prevent overflow. Likewise, if detritus, such as gravel and stones, be carried into rivers at flood seasons more rapidly than it can be ground up by attrition and carried away, the deepening of channels will be retarded or arrested.

BODIES OF THE DROWNED DRIFT TO THE SHORE.

The situation in which the bodies of persons drowned in rivers are found also bears evidence to an outward movement of the water at the bottom of streams. Such bodies, no matter in what part of the river the drowning may have occurred, are found almost invariably at the banks. If an outward underflow is conceded, this is readily explained. When a person is drowned, the body, being of slightly greater specific gravity than water, sinks to the bottom. After an uncertain lapse of time decomposition sets in and the body is made specifically lighter by the gases generated, and when it reaches the exact specific weight of the water it begins to drift. Usually it retains this exact specific weight long enough to reach the shore, where it is held by the attraction of the bank until discovered.

RISING OF STREAMS BELOW DAMS.

To the orderly movement of water in channels as herein set forth is due in great measure the capacity

of streams to carry large quantities of water in proportion to the size of their channels, and also the
more rapid rise of streams below than above falls.
Every one has remarked the much greater rapidity
with which streams during floods rise below than
above obstructions. In such cases the breaking up
of the double spiral confuses and checks the water
below the obstruction, while that above moves on
undisturbed.

As an instance of this action may be cited the
fact that in the Ohio River at Louisville the fall of
the bed from First to Thirtieth Street, a distance of
about three miles, is at the lowest stage 27 feet.
At a 2.5 foot stage the fall is 26.6 feet; at a 7.5 foot
stage the fall is 21.3 feet; while at a 46.7 foot stage
the fall is only 1.6 feet within the limits given.
Owing to the breaking up in this, as in all similar
cases, of the orderly double spiral and the conse-
quent confusion, the fall is stretched out over a long
extent of river.

It is true the greater facility of discharge just
above the falls has something to do with the results,
but a comparison of the rise at the head of the falls
with that at points unaffected will prove that the
difference is mainly due to the cause ascribed.

In 1883 at Davis Bend, twenty miles above New
Orleans, a crevasse occurred, permitting about one
third of the water in the Mississippi River to escape
across the country to the Gulf of Mexico. The distance to the sea level at this point is, on an average,
about twenty miles, with a total fall of about nine
feet. Now, such was the hindrance to the flow of

water, due to the want of a rightly proportioned
channel, that the water spread out over a territory
extending up and down the river about eighty miles
in length. At places the water on the slope running
back to the Gulf was from ten to twelve feet deep,
and almost as still as a lake.

Some of the New Orleans papers suggested that
the grass of the swamps and the water plants must
have dammed up the water, so much was it backed
up and so slow its movement. And yet the Mississippi River, with its orderly current, carried the
remaining two thirds of the water to the Gulf of
Mexico, a distance of one hundred and twenty-seven
miles, at a rate of five miles an hour, and with a
normal fall of less than two inches to the mile.

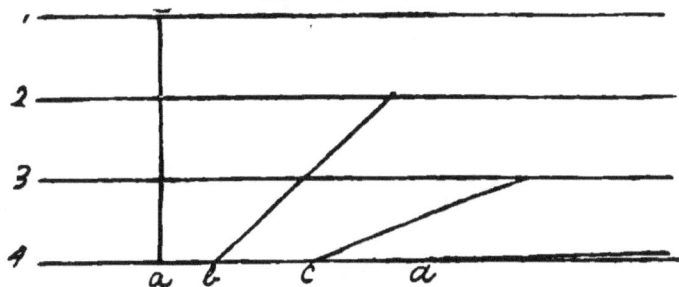

Fig. 3.

DEDUCTIVE EVIDENCE.

With the foregoing presentation of the argument
reached by induction, and before entering into a
detail of direct experiments bearing on the question,
some further support may be added, derived from
evidence based on deduction.

Let us, then, conceive a stream to consist of perpendicular columns of molecules (Fig. 3), and that

a row of these columns extending longitudinally in the axis of the stream, reaching from the bottom to the surface, is taken for observation. Now, since the lower end of all these columns is known to be retarded, and also more retarded than the top, it follows that in time they will all come to lean down the stream, and in so doing will not be able to reach the surface at a level which the stream is required to maintain in order to enable it to carry all the water that it must convey.

At *a*, for instance, the top of the column will reach to the surface at 1, at *b* it can reach only to 2, at *c* it will be down to 3 ; and by the time the column arrives at *d* it will, like all the rest in turn, at 4 lie flat on the bottom.

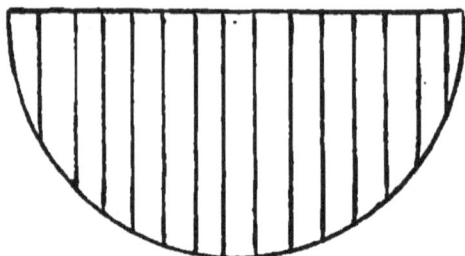

Fig. 4.

Thus, unless by some mechanism as yet unsuggested the columns can receive additions to their length, enabling them to reach the surface at its required level, the stream must eventually fail and disappear from very flatness.

On the other hand, let us conceive the stream to be formed of perpendicular columns of molecules, as before, and take a row of columns in cross section, extending from the bottom to the surface.

Now, as the stream flows on the column of particles next the bank will be retarded, and that progressively, and the result must be that it will eventually come to an absolute standstill. When the motion of the first has been arrested, the next within must have the same experience, and so on until the motion of all has been arrested, and the stream fails and disappears from very narrowness. If, however, a mechanism is provided for lengthening out the · columns in the first example, or of substituting more swiftly moving ones in the second, the stream may continue on without diminution in volume.

THE LESSON OF EXPERIMENTS.

When this theory has been presented from time to time to men from whom one might have expected a helpful judgment, a quite uniform reply has been, ''Prove it by experiments.'' And yet when one considers how many volumes of reports of experiments have been given out which were made under most favorable circumstances, and what an infinitely greater number have been made and have not been published, and all without even approaching a settlement of the question, it reconciles one to the contemptuous replies one often gets from men enviably favored in their positions for forming an opinion on this question. It does, indeed, seem like wanton presumption to undertake to thresh over the old straw and venture to repeat test experiments on this line. So, before I give results of my own experiments I may therefore be allowed to fortify my position by citing experiments

made by two eminent men already mentioned, viz.,
Professor James Thomson, the brother of Lord Kelvin, and Major Allan Cunningham.

EXPERIMENTS OF PROFESSOR THOMSON.

In the Transactions of the British Association for
the Advancement of Science, held in Glasgow in
1876, we find the following on page 31, it being a
part of a continued report of Professor Thomson on
fluids: "The chief view," says the report, "now
experimentally proved, was that the water in turning the bend exerts centrifugal force, but that a thin
lamina of the water at the bottom or in close proximity to the bed of the river is retarded by friction
at the river bed, and so exerts less centrifugal force
than do like portions of the great body of water
flowing over it in less close proximity to the river
bed. Consequently the bottom layer flows inward
obliquely across the channel toward the inner bank
and rises up in a retarded condition between the
inner bank and the rapidly flowing water and pro-
tects the inner bank from the scour, and brings with
it sand and other detritus from the bottom, which it
deposits along the river bank. The apparatus
showed a small river about eight inches wide and an
inch deep, flowing around a bend and exhibiting
very completely the phenomena which had been
anticipated."

MAJOR CUNNINGHAM'S CONCLUSION.

And this from Nature (Vol. XXV, page 1), in a
review by the editor of a two-volume report by

Major Cunningham, giving the result of many thousands of experiments made by him on the Ganges canal in India : " His own float observations show that near the edge of a stream there is *a persistent flow of the water at and near the surface from the edge toward the center.*" And, again, " The motion of water at each point varies in magnitude and direction from instant to instant."

Now, a little amendment which can hardly be with reason objected to, and these two experiments put together, prove the contention of this essay. In the first place, the outward movement at the bottom on the inside of the bend of a stream, as exhibited by Prof. Thomson, is the one most easily obtained in experiments.

As stated by him, the centrifugal force on the outside of bends carries the current to the outer bank. The outgoing undercurrent, then, is made to travel twice as far as it must do in a straight stream, and both its transverse direction and its retardation become very marked and distinct, and, consequently, easy to measure. But these outward undercurrents existed on both sides of the stream before this bending took place, only in a less degree. We will assume, then, that Professor Thomson's experiment has established the outward movement at the bottom.

Major Cunningham, as we have seen, as a result of many thousands of hydraulic experiments made on the Ganges canal in India under government supervision, and reported in his official capacity, declares that, "near the edge of a stream there is a persistent flow of water at and near the surface from the edges toward the center."

Now, if we reflect that, owing to the much greater speed of the water as it nears the center, its longitudinal motion greatly increases in proportion to its transverse motion, we can easily see how it could happen that the transverse motion was more likely to be and was overlooked by Major Cunningham. We then have Professor Thomson's experiments, made in the presence of the British Association for the Advancement of Science, establishing the fact of the outer undercurrent and Major Cunningham's official report affirming the inward flow from both sides, at and near the surface. With inward flow conceded at the top, and outward flow at the bottom, the other two necessarily follow

My own experiments do little more than simply bear out the others mentioned. As for the inward flow of the surface water, by throwing sticks into natural streams mainly, but often in artificial canals, I have observed it times without number through experiments purposely repeated for more than twenty years.

The outward flow at the bottom is not so easy to demonstrate experimentally. That tendency of the motion of water at each point to vary in magnitude and direction from instant to instant, as described by Major Cunningham, is only too much in evidence.

In such artificial canals as are ordinarily used for experiment, no satisfactory result has been obtained by me from allowing free material to be carried along by the water. The general trend of the water's motion may, however, be obtained by attaching substances of near the specific gravity of the water to long cords of fine thread, and then fastening these so that they will occupy the middle of the canals or sides, as one may choose. The total result of such experiments has been such as to bear out the conclusions drawn from so many other sources.

Many of the foregoing explanations have been based upon the assumption of the existence of channels. But it is evident that if the principles here presented are true — if the true theory of streams has been discovered—no channel formation could have taken place except in the way described.

In that case, the low lands of the earth must all have become a dreary expanse of seething, poisonous swamp and marsh, and the whole surface of the earth must have retained nearly the form with which it rose from the sea.

The fact that water becomes lighter as it freezes, and that winter covers the rivers with ice instead of filling up and obliterating their channels, has been considered a matter so vital in its bearing on human welfare that it has been widely advanced as an evidence of special providential care on the part of the Creator. Yet even without this provision we should still have left to us, undisturbed by winter's cold, all the rivers of the tropics and the adjacent parts of the temperate zones.

ESSENTIAL TO EXISTENCE OF HUMAN LIFE.

But in the absence of the principle we are considering, the maintenance of human life on earth had been scarcely possible, and its evolution abso-

lutely so.   Only in a few elevated localities could
any dwelling-place for human beings have been
found, and that necessarily of the most inhospitable
character.

There could have been none of the pleasing
alternation of hill and valley, of swell and swale, of
champlain and canyon, of rolling prairie and fluted
mountain slope that in endless variety have endowed
the earth with so much of charm and beauty.

First, the tablelands had to be notched by the
channels of streams whose existence was rendered
possible only by the concentration of the force of the
water's flow, and then the ceaseless chiseling of wind
and rain, and heat and cold, shaped the rugged
banks into graceful slopes and embellished the
mountain sides with a marvelous wealth of sculpt-
ured architecture.   Even the utility and attractive-
ness of the ocean are immeasurably enhanced by
this power of channel forming possessed by flowing
water ; for the harbors whose protecting quiet has
invited the great cities to gather about their shores
and share the fruits of commerce and travel have
nearly all been constituted by the subsidence of areas
of land into which valleys and canyons had been
carved by the rivers.   In view, then, of the beauty
and grandeur of rivers and their intimate relationship with whatever in the history and experience of
man he most may contemplate with satisfaction and
pride, it must be owned there comes a lively pleasure with the hope that hereafter, as long as men
shall find delight in the varied landscape, as long as
they shall revel in life and health, as long as glad

eyes shall be mirrored in glad waters, a new thought will have been awakened in every dreamy, azure depth, and a new note in the mellow music of every rippling, murmuring brook.

GLACIERS AND AIR-STREAMS.

The principles of the double spiral apply not only to the motion of water, but to that of all liquids and all gases, and to that of all viscous substances as well, when they are moving in channels. In the movements of glaciers the appearances, owing to the peculiar conditions, are to a large extent misleading, and experimental proof is not so easy to obtain as in the flowing of water. When a stream of water has developed into the double spiral, the small viscosity of the fluid allows the momentum of the parts at the surface to have nearly full play ; hence the motion of the upper stratum is rapid to the very center of the stream, where, as we have seen, the water is even heaped up into a ridge, and on its surface broken into wrinkles. If, however, the banks of a stream of water are very gently sloped, the motion of the cylinder does not extend quite to the margin, but leaves a limited extent of water in that situation but very slightly affected, and really forming almost as much a part of the channel as of the stream.

DIMINISHED MOVEMENT AT EDGE OF GLACIERS.

In glaciers this condition at the margins is much magnified, and even along the middle of the glacier there is a strip more or less broad that has no trans-

verse but only a longitudinal motion. If, as cur-
rently held by scientists, the viscosity in the ice
river is due to the pressure of the overlying mass on
the parts beneath, it can not extend unimpaired to
the surface, though it will be somewhat helped out
there by the heat of the sun. In addition to the
absence of pressure at the top, the surface in the
middle and at the margins is protected from the
sun's rays by the material forming the central and
lateral moraines, and is thus made more rigid than it
would otherwise be.

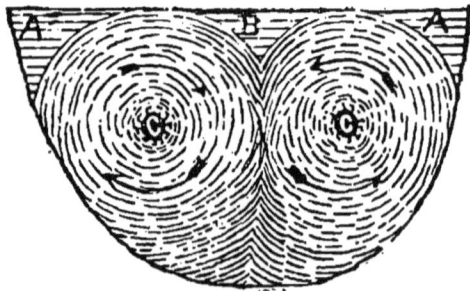

Fig. 5.

TRANSVERSE SECTION OF A GLACIER.
A—Unaffected ice at margin of glacier.
B—Medium part unaffected by lateral motion.
C—Center of cylinder.
The arrows show the direction of spiral motion.

The two cylinders forming the active part of the
glacier will not, therefore, extend to the surface with
full power, as a rule, so that a more or less extensive
mass of surface ice will be left nearly free from lateral motion as in the illustration (Fig. 5).

The inward movement of the upper surface of
the cylinder will, however, drag the less viscous
surface ice obliquely downward and inward toward

the middle, and innumerable fractures will therefore be made in the surface ice as it is pulled upon obliquely in this way.

ROCKS RISE TO THE SURFACE.

Rocks and other material dropped into the fissures in glaciers have been observed to rise to the surface at the edges, a fact which, for such instances, at least, would go to show that the ice rises at the margins. Doubtless the stones of the central moraines, when once sunk beneath the surface, go on to the bottom and move outward, to be lifted up again perhaps at the edge of the glacier. The fact admitted by all authorities, that the motion of glaciers at the sides and bottom is retarded, necessitates the conclusion that the double spiral motion obtains also in them. The retardation must be progressive. The part, then, of the glacier in contact with the sides must very soon under progressive retardation become stationary, and if its place in the moving column be not taken by other and faster moving portions, motion in the whole glacier would finally of necessity cease. For, as shown before, when the outer part became stationary from friction against the channel, the parts next within would become stationary from friction against these, and so on until the whole mass should be brought to a standstill.

It is not probable, however, that the double spiral action in any glacier was ever active enough to inaugurate a channel. Water, doubtless, first cut all the channels which now form the beds of glaciers.

STREAMS OF ATMOSPHERE.

The application of the principle of the double spiral to the movements of the atmosphere is quite obvious, but owing to the small amount of friction among particles of atmosphere, the question is embarrassed with infinite complications. Furthermore, it is only occasionally that conditions favor the manifestation of this motion.

On a small scale it is often exhibited in connection with summer showers occurring in a country of broken surface. It is a popular belief that streams of water attract rain-clouds, and doubtless territory in the vicinity of streams is favored with more than its share of summer showers.

It is not, however, that streams attract the rains, but the winds, moving in line with the extended depressions in which the streams move, take the double spiral form, and the clouds are drawn to the middle just as drift is drawn into the middle of a river.

BLIZZARDS AND TEXAS NORTHERS.

When a vast movement of the atmosphere takes place along the sides of mountain ranges this feature is brought out on a large scale. Thus, when areas of high pressure compel a great movement of the atmosphere southward along the Rocky Mountains, the mountains on the west side of it act as one bank of an enormous river.

The friction of the air against the mountains causes it to rise and climb up their sides, and just as we found the water in a river settling down in the middle to replace that moving out at the bottom, so the air from the colder regions above, which has out-traveled the lower strata and therefore come from farther to the north, settles down to take the place of the air moving up the mountain sides. The western blizzard and above all the Texas norther are in large measure a descending part of a spiral, determined by friction of the winds against the surface of the eastern slope of the Rocky Mountains.

THE GULF STREAM.

The so-called gulf streams, because first observed flowing into and out of the Gulf of Mexico, are not of the same nature as rivers, though the current is well marked off in a part of their course from the mass of the ocean. The power which propels these vast currents is in the main the same as that which supplies energy for the interchanging of the polar and equatorial atmosphere. It is derived from the difference of density between the water near the equator and that toward the poles, arising from a difference of temperature.

A slight influence is probably also exerted by the preponderance of evaporation in the tropics, the evaporated water being carried toward the poles and precipitated.

The water thus precipitated must of course join in the flow toward the equator. But if there were no evaporation, the water would expand much more under the equator than it now does, and the pole-ward overflow would be so much greater that it is

questionable whether evaporation does not detract
from rather than contribute to the activity of the
polar and equatorial interchange.

The direction of these ocean currents, however,
as we find them in the tropics, concentrated into
streams, is doubtless determined by the westerly-
blowing tradewinds, as is also a large part of their
movement.

It is not possible, however, that the energy that
drives these vast flows of water far even into the
polar seas can be the momentum imparted to them
by the converging tradewinds.

After they have turned poleward and escaped
from the region of the tradewinds, they pass, in so
far as direction is concerned, mainly under the control of the rotary movement of the earth.

But the momentum previously obtained doubtless persists for a time as a propelling force. The
most active cause of movement, as already stated,
is the indirect operation of gravity pulling them
down the actual incline of the ocean plane, due to
the expansion of the equatorial waters.

Their return, on the other hand, is largely along
the potential incline of the cold waters beneath,
which extends toward the equator.

When these streams are left almost wholly to
the influence of gravity, as when they turn and start
back toward the equator, they break up and lose
their identity.

But as long as they preserve the proportions of
a river, as on the coast of Florida or Japan, for
instance, they are subject to the same laws that
govern rivers.

Yet, owing to the greatly limited friction between the ocean stream and its watery walls, as compared to that between a river and its banks, the double-spiral must be much less conspicuous.

This character of motion must also be supplemented to a considerable extent in the currents which move from east to west along the equator, since the wind which mainly produces them is constantly driving the water of the surface obliquely inward from either side. In this case the double-spiral is promoted by friction at the surface instead of the bottom, as in rivers.

PRACTICAL APPLICATION.

The practical application of the principles considered in this essay may facilitate the discovery of general formulas for calculating the capacity of canals intended for conveying water at varying degrees of inclination, and of methods for keeping irrigating canals free from boulders at the point where water is first led into them from rivers. The discovery may also throw light upon the question of the best means for improving and controlling rivers.

The question of this nature which for the people of this country surpasses in interest and importance is that of the improvement and control of the Mississippi River, although nearly every country has similar occasion for its special application.

The problem of the Mississippi involves the consideration of three important measures, in regard to which there is much diversity of opinion among engineers.

These measures are the construction of jetties, the erection of levees, and the maintenance of outlets. The jetty system, as now in operation at one of the outlets of the Mississippi, stands approved throughout by the principle of the double spiral ; for it aims at contracting the width of water flowing over bars so as to secure the proper relation between the width of a channel and its depth, where otherwise the tendency would be for the stream to widen at the expense of its depth, or even to divide and form a number of channels. The limit of depth to which any body of water flowing through jetty walls can attain will be the point at which the accelerated mass of water sinking down in the middle no longer strikes the bottom with sufficient force for purposes of erosion — a problem which may some time be exactly formulated.

All the jettying, however, that it can ever be in the power of any government to do, while it may improve navigation, can never materially affect the problem presented by the great river, as relating to its control.

The question of levees is a matter which presents greater financial than scientific difficulties. The levee system likewise is based upon nature's own plan, for we have seen that a river flowing into a shallow body of water will, if it carry sufficient silt, build up at first a bed for itself, and then throw up levees on its bank so as greatly to limit if not eventually arrest its own overflow.

But nature contemplates a continuous levee and a limited overflow at any one point, and not the

building up of a levee so as to accumulate a large
head of water, to be allowed to pour through at
a few points and open deep outlets to the sea. The
width of the river must be proportioned to its depth ;
and since cultivation of the soil at the headwaters
of the Mississippi and the restraint of the shallow
overflow which once took place over a long line of
bank have greatly increased the volume of water
to be carried by the river in the lower part of its
course in times of flood as well as prevented the
ejection of silt, the levees along the corresponding
part of the stream must be constructed far enough
back from the banks to allow the channel to widen
and the river to re-establish its regimen.

In the absence of any disturbing cause, such as
rock in the bottom of the channel or excessive
amounts of detritus, the slope of a river will, as we
have seen, be regulated from the point of the outflow at the sea ; that is, the slope from that point
will be no more than is necessary to admit of the
flow of the water. As the sea is pushed farther
away by silt deposit, the river bed must be elevated
along its whole line above in order to preserve the
proper incline. This will, of course, produce increased overflow of the banks until they can be sufficiently built up to prevent it.

Without, then, considering the question of outlets,
since we know that the Mississippi has been steadily
encroaching on the Gulf with its delta, there has
been no time in its history when it did not overflow
and send its waters off to the Gulf by lateral routes
above New Orleans. All of the flood water of the

Mississippi never did pass New Orleans. Other proof would hardly be necessary to demonstrate that the levees below the mouth of the Red River are neither high enough nor located sufficiently far from the banks of the river, if the conveyance of the entire body of water discharged by the Mississippi is contemplated. The principal if not the insuperable difficulty of constructing such a system of levees as the conditions require is the enormous pecuniary outlay involved in the undertaking.

The propriety of constructing outlets which shall lead off a part of the flood water of the Mississippi at some point above the mouth has excited much controversy. But consistently with the principles herein advocated, outlets would prove futile or harmful according to their extent and the objects they might be expected to accomplish. The three great outlets above New Orleans, namely, Plaquemine, Lafourche, and Atchafalaya, which were most likely formed under geological conditions different from those which now obtain, had nearly or completely closed up at their sources when Louisiana was first settled and when the building of levees was first begun.*

* Humphreys and Abbott, in their report on the Mississippi, contend that these outlets never were mouths of the Mississippi, but have been formed by overflows. They also maintain that the present great muddy Mississippi is less than five thousand years old, and never had mouths above the present outlets at the passes. They may be correct as to the age of the Mississippi in its present form, but these bayous, with countless others which can yet be traced in the deltas, were doubtless made by the comparatively clear and small stream which formerly occupied the line of the present Mississippi; and when the latter broke through from the great inland sea that its waters then formed, it first filled them up at the head with silt as it advanced to the ocean, and subsequently cut them out again under the great pressure due to increased head of water produced by levees. Such bayous have been made through the "tule" lands of the Sacramento and San Joaquin; every delta river in the world has them, and there seems but one way in which they can be formed.

The first two, through the impetus and volume given them, have been enlarged into considerable streams, while the third, the Atchafalaya, seriously threatens to become the main channel of the Mississippi. A complete maintenance of levees along that part of the Mississippi bordering on the Atchafalaya basin, leaving the outlet open as now, must in a few years result in turning the Mississippi into the Gulf by way of the Atchafalaya, thus completely shutting New Orleans off from river navigation, and leaving it on an arm of the sea.

The fear that outlets may cause the main channel of the river at points just below them to fill up with silt need not be seriously entertained until the outlet shall have become the main channel. It has been shown that the water sinks downward in the middle of streams and moves out at the bottom with constantly diminished speed, to be almost completely arrested at the banks. Deposits of silt, then, should first be expected at the banks below outlets, and not in the middle of the stream; at all events the deposit at the sides and middle would be proportional after the stream had recovered from the disturbance due to the opening of the outlet. As regards the Lake Borgne outlet below New Orleans, first officially suggested by Humphreys and Abbott, and since warmly advocated by many, it is not easy to see how it could prove of any considerable efficiency, or be maintained except at great cost, unless it should first have become the main channel of the river. On account of the narrowness and shallowness of the outlet by this route, the outflowing water would be so retarded that an extensive deposit of silt would take place at the lower part of the lake at every overflow, and this would continue to increase until the outlet should be practically closed. In the mean time the transverse under-current of the river at flood would throw silt into it at its head, so as to necessitate each returning season a large outlay for its removal.

But if, on the other hand, the scouring should progress, and one of those rare occurences—the turning of a great river from its channel—should take place, the lifetime of generations might be required for the new channel to be bored out and leveed up by the current along the new route so as to make it available for navigation. Such a change, though possible, is not likely under the conditions existing at that point.

RED RIVER CHANGED ITS ROUTE.

Owing mainly to the influence of the double spiral, water moves much more easily and economically in large streams than in small ones. Thus Red River chose to give up its route of one hundred and eighty miles through the Teche to the Gulf, and now travels three hundred and sixty miles to the same destination, for the reason that it proved to be a measure of economy to join forces and to travel the greater distance in company with the Mississippi than to traverse the shorter distance alone and in its own smaller channel.

Still, when a delta is pushed far out into the sea and a river is compelled to flow on a narrow ridge, with the loss of incline incident to such a condition, the time must come when it will break from its channel and seek the sea at some nearer point to the right or to the left. But when that time is to arrive in the history of the Mississippi remains for the future to disclose.

www.ingramcontent.com/pod-product-compliance
Lightning Source LLC
Chambersburg PA
CBHW030832270326
41928CB00007B/1017